Parasites, People, and Places
Essays on Field Parasitology

Professor **Gerald W. Esch** is one of the world's leading ecological parasitologists. Here, he presents a series of essays on classic examples of field parasitology. The essays focus on the significance of the work and its contribution to the field, but also on the people and particularly the sites at which the work took place. Taken together, the essays represent a beautifully written account of the development of an entire field of scientific endeavor spanning a period of 50 years or more. The essays are not meant to be academic in a scientific sense, but there is a great deal of science in them. The book will be of great value to all parasitologists and ecologists, but also to anyone interested in how biological field work is carried out and how it contributes to a greater understanding of the natural world.

Parasites, People, and Places

Essays on Field Parasitology

GERALD W. ESCH

Charles M. Allen
Professor of Biology
Department of Biology
Wake Forest University
Winston-Salem,
North Carolina, USA

CAMBRIDGE
UNIVERSITY PRESS

Shaftesbury Road, Cambridge CB2 8EA, United Kingdom

One Liberty Plaza, 20th Floor, New York, NY 10006, USA

477 Williamstown Road, Port Melbourne, VIC 3207, Australia

314–321, 3rd Floor, Plot 3, Splendor Forum, Jasola District Centre, New Delhi – 110025, India

103 Penang Road, #05–06/07, Visioncrest Commercial, Singapore 238467

Cambridge University Press is part of Cambridge University Press & Assessment, a department of the University of Cambridge.

We share the University's mission to contribute to society through the pursuit of education, learning and research at the highest international levels of excellence.

www.cambridge.org
Information on this title: www.cambridge.org/9780521815499

© Cambridge University Press & Assessment 2004

First published 2004

A catalogue record for this publication is available from the British Library

Library of Congress Cataloging-in-Publication data
Esch, GeraldW.
Parasites, people, and places : essays on field parasitology / GeraldW. Esch.
    p.; cm.
Includes bibliographical references and index.
ISBN 0 521 81549 5   ISBN 0 521 89457 3 (paperback)
1. Parasitology.   2. Parasites.   I. Title
QL757.E79 204
591.7′857–dc21   2003051529

ISBN   978-0-521-81549-9   Hardback

Contents

Preface

What is an epigram? A dwarfish whole. Its body brevity, and wit its soul.

SAMUEL TAYLOR COLERIDGE, *AN EPIGRAM* (1802)

According to my *Random House Dictionary*, an essay "is a short literary commentary on a specific topic, usually in prose and generally analytic, speculative, or interpretive." Field sites and biological stations are geological and physical entities, just that, really nothing more. Most of them, of course, are beautiful to look at and wonderful places to visit and work, in the summer time, especially. The woods or prairie that might be present, a nearby lake or pond, usually provide a marvelous pastoral perspective. However, the real story of any field station or field study site is actually more associated with the people who go there to do their "thing." Accordingly, this book and these essays are mostly about these people and their research, or their teaching, at field stations or a special field site. I was primarily interested in why they chose these locations, what drew them into the work that was done there, and what impact it has had on the way in which we view the ecology or epidemiology of parasitic organisms.

A couple of points should be made at the outset. First, when I registered in the fall semester of my senior year at Colorado College in 1957, I wanted to take a couple of history courses, but had signed up for just one, in zoology, my major. Approval had to be obtained from the Department Chairman, Professor Robert M. (Doc) Stabler, who also was my advisor. When I went in to see him, he took the registration card from me and studied it, carefully, like he was thinking of something he wanted to say. After the proverbial "pregnant pause," he asked, "Are you going to be a history major or a zoology major?" I responded, wisely, "Sir, a zoology major." "Very well," he said, "you need to take my parasitology course," which I did and, of course, became a parasitologist as a result. However, I never lost my fascination for history. This interest will become apparent when you read these essays because there is a clear historical component in most of them. I make no apology for including this information because I think I have made it relevant to the stories I am telling in each case; at least I have made an effort to do it this way. You will have to be the judge of whether I was successful or not. Second, in a couple of essays, I have rambled a little bit. I thought about removing some of this "fluff," but could not persuade myself to do so. I think the fluff adds some "flavor" and brings a personal tone that an essay should have. Again, you must be the judge of whether I have succeeded.

The year 1914 marks the beginning of parasitology research at the University of Michigan Biological Station by William Walter Cort, and, as near as I can determine, it also represents the beginning of field parasitology research, as we presently know it, in North America. This is not to say that parasitology research had not been done in North America prior to 1914, because it had. Indeed, Joseph Leidy (1823–91) is described by none other than Henry Baldwin Ward as the "father" of American parasitology. Some refer to Ward in this regard. If the latter are correct, then I guess we could say that Leidy is the "grandfather," though not of the same academic bloodline as Ward. Several Canadians, e.g., William Osler, William MacCallum, and Ramsay Wright, also worked in North American parasitology during the late nineteenth and early twentieth centuries, but none of them was involved in what can be considered in today's terms as "field parasitology," and neither was Joseph Leidy. I am not as familiar with historical parasitology in other parts of the world, although it is clear that a great deal of life-cycle work was accomplished in Europe, especially in the last half of the nineteenth century. Substantive discoveries were also made in Africa and the Indian subcontinent about the same time, mostly dealing with parasites of humans and domesticated animals.

I have begun these chapters with a Prologue in which I describe some of my own experiences regarding the ecological aspects of host–parasite interactions, and what attracted me to this area of research many years ago. You will see that what I have written in the Prologue and in several of the other chapters will appear somewhat autobiographical, but the book is not intended to be an autobiography. A great deal of what I have to say is based on my personal knowledge and interaction with a number of colleagues who have been "players" in ecological parasitology over the years. Regrettably, I have omitted reference to a number of the outstanding field parasitologists of the earlier and modern eras in Europe (including Scandinavia), Australia, New Zealand, Russia, and even the western hemisphere. And, I did not write about several of the prominent field stations of the day, like the ones at Woods Hole, Friday Harbor, the Gulf Coast Laboratory, and Lake Itasca. When I suggested this book to my Commissioning Editor, Ward Cooper, at Cambridge University Press, he sent the proposal to several parasitologists in various parts of the world. Most of them liked the idea. A couple of them, however, wondered why I did not make reference to one or other of their favorite field sites, or biological stations, or parasitologists? I could have replied that I was limited by space, but this would have been only partly true. I think it is more accurate to say that a lack of experience or first-hand information regarding their favorite sites, or specific knowledge of the people who did the work there, were the real limiting factors in what I have attempted to include here.

There are not many biological stations devoted strictly to the study of parasite epidemiology. In fact, the only one that I can think of in the western hemisphere is the now-defunct Gorgas Memorial Laboratory in Panama. Most epidemiological work is conducted in the field at sites where one encounters a given parasite of interest. For the most part, this would be in small villages or communities, mostly in the rural areas of developing countries throughout the world. As a result, it is difficult to identify a special site and, therefore, to deal with this particular topic in the same way that I have the others. So, the approach in the chapter on epidemiology is somewhat different. Moreover, because of the vast array of parasitic diseases and the problems they create, I had to restrict myself to just a few examples, ones that were of interest to me, about which I knew the most, or about which I was able to persuade a friend to contribute something special.

So, this gives you some idea about what I wanted to do with this book, and why. I sincerely hope that I have succeeded.

Acknowledgments

A large number of folks gave their time, or documents, to help me in this effort and I thank all of them, beginning with Katarzyna, the daughter of Wincenty Wisniewski, who sent me her father's short autobiography. What an interesting, and dangerous, life this man led! Jerzy Rokicki, Elzbieta Zbikowska, and Thadeus Graczyk, all proud Poles, either read, or in some way contributed to, the chapter dealing with Lake Druzno and Wisniewski's work there. I appreciate their assistance. I was invited to spend several hours with Mrs. Helen Brackett and Mrs. Margaret (Peggy) Olivier, daughters of Will Cort, one my early "heroes" in parasitology. They were both truly fascinating ladies and I sincerely thank them for all their help and encouragement. Dick Seed accompanied me on one of my visits with the Cort daughters and helped with my "interrogation." He later read the Cort chapter and the others as well, and commented on them for me. I appreciate the time he took to do it. Brent Nickol kindly sat with me for an hour or so in Albuquerque, New Mexico, and answered questions about himself, his career, and his relationship with the Cedar Point Biological Station. John Janovy, Jr. allowed me to spend a couple of days with him and his students out at Cedar Point where he teaches his field parasitology course. That was such a great experience for me! His wife Karen listened as John and I kibitzed about our experiences at the University of Oklahoma some 40 years ago. She also provided me with some information on the development of what was to become the Cedar Point Biological Station. What gracious people they are.

I thank my friend of more than 30 years, Clive Kennedy, for talking with me several times in his office at the University of Exeter. We have had many discussions about Slapton Ley and his work at this unique field site. He was also kind enough to read a couple of drafts of the essay dealing with

Slapton Ley and offered a number of excellent suggestions for improving it. Among other things, he gently set me straight regarding Mary Stuart, Queen of Scots, and her relationship with Elizabeth I.

Toward the end of my writing, I received a fascinating little volume, entitled *Wendell Krull: Trematodes and Naturalists*, written by S. A. Ewing and published by the College of Veterinary Medicine, Oklahoma State University, where Ewing holds the Wendell H. and Nellie G. Krull Chair of Veterinary Parasitology. This book was a wonderful source of information regarding a truly remarkable early parasitologist, Wendell Krull, and was used to "flesh out" several stories regarding a number of parasites on which he worked and people that he knew in that era.

One of the essays has a focus on Will Cort and the University of Michigan Biological Station (UMBS). I am grateful to the current Director of UMBS for providing me with two documents from which I obtained much useful information on how the Station got its start. One of these was a program from the semicentennial celebration held at UMBS in June of 1959. The other was a historical account of UMBS from 1909 to 1983, written by David M. Gates, who was Director of the station for many years.

Ward Cooper, my Commissioning Editor at Cambridge University Press, requires special acknowledgment. I have known Ward for a long time and we became good friends in the interim. He has been very supportive of the various projects in which I have been involved at Chapman and Hall and at Cambridge University Press during this period. Ward selected a number of people to read the prospectus for this book. Even though they were supposed to be anonymous, several have confessed to me that they critiqued the proposal and I have already thanked them personally for helping, but I want to thank them again here for their interest. To the others who are still unknown to me, I thank you too for your help and interest. All your suggestions were greatly appreciated.

When I was in Albuquerque, New Mexico for the annual meeting of the American Society of Parasitologists in 2001, I was persuaded by Tim Goater and David Cone that I had to include an essay on Canadian field parasitology. To help do this, David sent me a marvelous little book on the history of Canadian parasitology written by the really fine parasitologist, the late Murray Fallis. I also managed to secure some good information from my old friend, Sherwin Desser, who helped me out with field research at Algonquin Park, north of Toronto. Sadly, I learned of the passing of Roy C. Anderson from Sherwin, who then volunteered a document Roy had sent to him about 3 months before Roy's death. It was full of invaluable information regarding his research experiences in Algonquin Park before he left the Ontario Research Foundation and went to the University of Guelph.

Dave Marcogliese, Peter Hotez, Dan Brooks, Darwin Murrell, and Mick Burt also corresponded with me about their field experiences, in some cases extensively, and I thank all of them. Through Darwin, I was able to obtain some photos from John Cross having to do with "the mystery disease of Pudoc", and I wish to thank John as well.

My talented daughter, Lisa, painted the cover of the book for me from a photograph I took on a visit to Clive Kennedy's lab in 1987. The painting now graces a wall in my office here at Wake Forest. Professor Emeritus of Biology at Wake Forest, Charles M. Allen, photographed the painting, and this is what actually appears on the cover of the book, so I want to give him a special "thank you."

I thank Vickie Hennings and Herman Eure for their help in so many different ways. Joel Fellis, one of my current graduate students, kindly prepared the figures showing the thermal and dissolved oxygen profiles of oligotrophic and eutrophic lakes and I appreciate his very good work. He and Nick Negovetich, another of my current graduate students, were also generous of their time in reading one of the last drafts of the book and offering a number of excellent comments. Mike Barger, who received his Ph.D. with me in the spring of 2001, provided some good commentary on parts of an earlier draft.

The photographs came from a number of sources. In a few cases, the photographer is unknown to me. I thank them too. Whereas I have identified each photo with the name of the person who either gave it to me or who took the photograph (if no name is attached, then I took it), I want to acknowledge all of them again here. Included are: Charles Allen, Al Bush, John Cross, Ann Esch, Joel Fellis, Jackie Fernandez, Tim Goater, Russ Hobbs, Mirna Huerta, Bob Kabata, Travis Knowles, Murray Lankester, Darwin Murrell, Margaret Olivier, Don Robinson, and Jerzy Rockicki.

Since I began directing graduate students here at Wake Forest in 1965, 36 have successfully made it through to either a Master's degree or a Ph.D. I am quite proud of these students because all of them have gone on to have successful careers. As I did in 1958 when I went to work with Professor J. Teague Self at the University of Oklahoma, most of them came to Wake Forest knowing of me only by reputation. I trust their hopes were realized. I know that mine were. Most of these students worked in the field with me or accompanied me to the Kellogg Biological Station when I taught during the summers from 1965 to 1974. Several of them were with me when I was actively engaged in research at the Savannah River Ecology Laboratory and a bunch of them worked out at Belews Lake or in Charlie's Pond. I especially hope they will appreciate why I am writing this book and know that they contributed to it in a variety of ways over the years.

One of these former students, and a very good friend, Tim Goater, took the time carefully, exhaustively, to go through the next-to-final draft of the book. He read it from cover to cover and provided me with an enormous list of comments, queries, and suggestions. I really owe him a great deal for his huge effort and contribution. It was invaluable in producing the final version.

Finally, I want to thank Ann, my wonderful wife of nearly 44 years. We have been through a lot together. I wish there could be another 44, but I don't expect that to happen. So, I'll take whatever is given us and enjoy the hell out of it. This book is dedicated to Ann and to our three wonderful children, Craig, Lisa, and Charlie. All three of them accompanied their mother and me to Gull Lake for 10 summers, to London in 1971–2 where they were forced to attend school in a "strange" (to them, at least) country, then to Aiken and the Savannah River Ecology Laboratory where we lived for a year. (Life in Aiken, a small southern town just north of Augusta, Georgia, was almost as bewildering, in some ways, to the children as it was in London!) Without my family and their support in all these travels and adventures, it wouldn't have worked.

Prologue

In my end is my beginning.

It's a poor sort of memory that only works backwards, the Queen remarked.

Back in the spring of 1958, I finished my undergraduate efforts at Colorado College where I had come under the influence of Professor Robert M. (Doc) Stabler and decided that I wanted to obtain a Ph.D. in zoology/ parasitology. So, I entered the graduate program at the University of Oklahoma where I was privileged to begin work on my Master's degree and then on my Ph.D. under the tutelage of Professor J. Teague Self. Whereas I was not too keen on it at the time, in the summer of 1960, I found myself at the Oklahoma University Biological Station (OUBS) down at Lake Texoma, a huge flood-control reservoir on the Red River, which separates Texas and Oklahoma. While an undergraduate student, I had done a lot of jackrabbit hunting on the high plains of eastern Colorado, searching for coenuri of *Taenia multiceps*, but OUBS was my first experience at a field station. It was not my last.

Most of the field stations in North America are located near lakes. I am not sure why, but perhaps next to an air-conditioned molecular biology laboratory, a lake is about the coolest place to spend a summer in the northern half of the world. In addition to basic biological research, most field stations are also involved in teaching. Back in those early days, parasitology courses were frequently taught at these stations, although that number has dwindled in recent years. A good field parasitology course requires students to do a substantial amount of necropsy work, which means that hosts of various sorts must be available for examination.

Being situated near a lake usually indicates that the first hosts to be looked at by field parasitology students are likely to be fish. Of course, we all know that the most enjoyable way of collecting fish is to take a favorite cane pole, a can of worms, and find a shady spot in the quiet cove of a small pond on a warm, sunny afternoon. Unfortunately, this requires time, and time for a student at a field station is in especially short supply. This is because students are generally required to collect x number of parasites, prepare a slide collection, identify the parasites, listen to lectures, take exams, eat, plus all the other mundane chores necessary to sustain one's self out of the classroom or lab on a day-by-day basis. Even finding time to sleep can be difficult on occasion. So, the quickest way to obtain fish is to set a gill net in the evening and retrieve it early the next morning. Although this may be the quick method, it is not always the most efficient way to collect fish for necropsy in a teaching lab. For example, the catch may be limited and insufficient to satisfy the needs of an entire class. Conversely, there may be too many fish. By too many, I mean that whatever one catches in a gill net must be necropsied immediately. It's not a good thing to throw away any vertebrate host, let alone gill-netted fish, once they have been brought into the lab for necropsy. Obviously fish could be frozen for later work, but really to learn about parasites, one needs to find and study them while they are still alive and moving. Accordingly, I can recall many days that first summer, setting gill nets in Lake Texoma from a small row boat, bobbing on 3-foot waves like a cork, getting the nets tangled, and receiving an occasional fin in the hand from a scrappy white bass or channel catfish. (By the way, never stick your fingers into the mouth of a bullhead or catfish – they have a very strong bite, and many, many sharp teeth!)

Of course, there are many ways to collect birds and mammals. If you are a good shot, a .22-calibre rifle will do the trick. If you are not a good shot, like me, then a .410 or 16-gauge shotgun should be used, except that whatever you might hit runs a good chance of being rather badly mangled as a result. A mist net set at dusk is really a good way to capture small birds, and snap traps work well with small mammals. Of course, frog "gigging" can be fun on a dark night at a local pond, although the "Tim Goater method" of capturing frogs is the best and most unique way of capturing these wily amphibians (see Chapter 3). Another time-honored procedure is to drive the country roads around the field station, searching for DOR (dead-on-the-road) carcasses at night, before the vultures or crows can get at them the next day. In many ways, this was perhaps the most precarious of our collecting methods down at OUBS, not because we were any danger of being hit by a speeding moonshiner (Oklahoma was still "dry" in those days) driving a souped-up, old truck. The problem was, we never knew just how long the DOR carcasses had been lying on the road. The Oklahoma sun in July can get mighty hot and there is nothing more internally combustible than the cecum of a ripe rabbit, for example. The potential for disaster is especially exacerbated when the necropsy of a DOR carcass is done immediately after breakfast. It is easy to see why the parasitologists at a field station are usually isolated, or, shall we say, shunned, by the rest of the biologists (especially the botanists) who are working there. There is definitely an aura associated with parasitology students, and their lab. Despite this guilt by association, we none the less collected all of the DOR carcasses we came across during those nighttime excursions. Well, that is, almost all of them. We consistently avoided skunks! Even young parasitologists-to-be had more sense than to haul something like that into the lab, unless it was late at night, and no one else was around (and that almost always meant the botanists). One had to be especially careful of DOR skunks because evidence of one of these critters in the lab is difficult to hide. The scent from a skunk has a tendency to settle on everything, seemingly for miles around, and then stay there. A bloated rabbit cecum is bad enough, but that smell is usually localized in the immediate vicinity of the lab – not skunks, though.

That summer also gave me my initial experience with collecting turtles and their parasites. Turtles are really beautiful animals and even more wonderful as hosts for parasites. The amount of time required to trap a turtle, however, is inversely proportional to the amount of time necessary to prepare a turtle for necropsy. The carapace must be removed, and that is not an easy job. We found the best way to catch a turtle that first summer was to set a basking trap in one of the ponds at the local fish hatchery in nearby Tischomingo. A basking trap is a simple contraption, consisting of a box frame made of 2 × 4s, covered with hardware cloth. One of its six sides is left open and the trap is then set in the pond with just enough above-water to invite a turtle to come and rest in the sun. When the turtle finishes warming, it leaves the perch and can go in one of two directions, meaning that 50% of them end up caught in the trap where they remain until pulled from the pond. That summer at OUBS, my partner in the trapping business was Jim McDaniel. I remember one day, especially. We found our traps filled with 40–50 turtles – real "pay dirt" for a field parasitology class – and we returned triumphantly to the lab prepared to share our spoils with the other students. Unfortunately, however, I was not to benefit from our huge success. Later that day, I became terribly ill. They thought at first that I had suffered a heat stroke, but one of the local physicians discovered I had acquired hepatitis. Although it was a mild case, I was quarantined with my family for about 10 days, until they were sure I was no longer infectious. Despite becoming ill on the day we had the great trapping success, turtles and their parasites have always held great fascination for me and, subsequently, for several of my students as well.

Another group of animals to which I was exposed that summer were snails. In many ways, these animals are about the best hosts with which to work, especially in these days of protocols, and Animal Care and Use Committees. There can be problems, however, particularly if one is trying to raise and maintain them for long periods in the lab. They really are quite demanding of one's time. In this regard, I am reminded of the experience of Miriam Rothschild, the wonderful "jumping-flea" lady. Prior to the outbreak of World War II, she had worked for 7 years on snail-trematode interactions at the Plymouth Marine Station on the southern coast of England. Her efforts were highly intense, so much so that she found herself tied to the field station and her aquaria for days and weeks at a time, without breaks. Just before the bombing of England began in 1940, she tells the story of going to the station director and asking if there wasn't some way to protect her precious animals. She was greatly concerned that her life's work would be lost in an air raid, but not to worry, she was told. Within a few days, however, her fears were justified when a night bombing raid by the German Luftwaffe destroyed virtually the entire laboratory, including her many years of work. Her immediate reaction was total rage at the loss of her laboratory snails and their parasites that she had so meticulously and tirelessly maintained. With her snails gone, however, she found herself suddenly untethered, and only then did she realize how much her life had become constrained, even dominated, by these small animals and the even smaller ones inside them. As she is quoted in an interview with *Scientific American* in May 1996, "Without realizing it, I had gradually become an appendage of my trematode life cycles." She is further quoted in S. A. Ewing's book (2001), *Wendell Krull: Trematodes and Naturalists*, "I packed my bags and left Plymouth, never to return as a research scientist. At the time I did not know that after World War II, butterflies and fields of flowers were to be exchanged for free-swimming cercariae and the turbulent Atlantic. I abandoned them with a touch of melancholy." Having spent some time with both trematodes and snails, I can see why she felt that way.

For any student who studies parasitology at a field station, the moment of truth comes in trying to find that first parasite. Imagine if you can, or recall if you have done it before, gazing down on your first carcass. It might be the gill-netted fish, a road kill, or a turtle captured in a basking trap. The animal must be opened, the organs separated, blood smears made, and the intestine slit open. This last act is what makes a "gut-scraping" parasitologist since one must literally scrape the gut, usually with a thumb, to dislodge certain parasites! The description of what is required to isolate, fix, stain, and then properly mount a helminth parasite would easily fill a chapter, but that is not my purpose here. Once the specimen is stained and mounted, then comes the really hard part, and that is the parasite's identification. One would think that when a beautiful helminth parasite is stained and mounted properly, everything should fall in place easily. But the problem is – for the "rookie" parasitology student at least – there are those things called dichotomous keys. This may appear to be a paradoxical statement since keys are supposed to make identification easy, but it usually isn't for anyone who has ever tried to use a key for the first time. Some say that experts are the only ones who can use them, and, in turn, they do not need them. There is some truth in this statement, to be sure.

For us that first summer, the rudiments of taxonomy were passed on from Professor Self, who directed us to the works of Satyu Yamaguti and *Systema Helminthum* (1958), the huge, multivolume series on parasitic worms of all sorts. Another invaluable source of information was a book that became my bible, and probably that of many field parasitology students of that era, *The Zoology of Tapeworms* (1952), written by a couple of Canadian parasitologists, Robert A. Wardle and James Archie McLeod. The cover of my copy of this book is now tattered, quite literally threadbare, and the binding has come apart, but I still have the one I purchased in March of 1960 when I became convinced that parasites were going to be my life's work. Learning field parasitology at Lake Texoma did not happen suddenly that summer but, finally, everything did come together with respect to my professional career. I wonder how many of us can look back and say, for sure, that was when I really became a parasitologist, not a very knowledgeable one at first, but you could none the less define yourself in a biological context.

Whereas I did biochemistry and physiology for my Ph.D. research, I was none the less still connected with the field since, throughout my

remaining Oklahoma days, I was forced to go outside the laboratory to secure adult and larval parasites for my biochemistry and physiology studies. I worked on the coenuri of *Taenia multiceps* and, at the time, the only
way these large bladderworms could be obtained was by shooting jackrabbits in the field. So, twice or three times a week in the third and fourth
summers of my graduate work, I would travel 90 miles west from Norman
to Fort Cobb, Oklahoma. There, I would be driven into nearby pastures
by a crusty old trapper named Lyle G. Rexroat who was employed by the
rodent and predator-control division of the US Department of Interior.
From the cab of his banged-around pick-up truck, I would shoot jackrabbits with my 16-gauge Mossburg. I almost felt like I was riding "shotgun"
on one of the overland stagecoaches as we bounced through the pastures
near the old frontier town of Fort Cobb.

The rabbits had no fear of his truck and the pickings were easy. They
were skinned in the field, the coenuri carefully removed and then packed
in physiological saline for a fast trip back to Norman in my tiny, and unair-
conditioned, Renault Dauphine. In those days, the trunk of a Renault was
in front of the car and the engine was in the rear. The trunk, for some reason, had a plugged hole at the bottom. I would pack the trunk with ice, and
load my jars of saline and coenuri, then trundle back to Norman with my
parasites. After the ice melted, simply pulling the plug allowed the water
to drain – a well-designed field vehicle was the old Renault!

That summer sticks in my mind primarily because of the fieldwork –
certainly not because of the enzymes assayed, or the long hours spent trying to set up and run the old Warburg manometers. Imagine running a
Warburg during an Oklahoma thunderstorm when the barometric pressure was changing the manometric readings more rapidly than the oxygen being consumed by the tapeworms. I should have known then that
my heart was not in the lab on those hot summer days, and especially later
when the two biochemists on my dissertation committee did not show the
slightest interest in the physiology or biochemistry research I so assidu-
ously attempted to undertake. When I defended my doctoral dissertation,
all they could ask about was how many of these rabbits were infected with
coenuri? Why was there a difference in infection prevalence between males
and females? Why were the coenuri found mainly in the hindquarters of
the rabbits? A biometrician on my committee kept asking me, why is there
such incredible variation in the numbers of coenuri in the jackrabbits? I'll
bet that Harry Crofton could have given him the answer, but I was not experienced enough to see what was going on with respect to the parasites'

Figure 1 The elegant Tudor-style home of W. K. Kellogg, the focal point of the W. K. Kellogg Biological Station of Michigan State University. The teaching laboratory for my field parasitology course was located on the second floor at the near-end of the building. (Photographer unknown.)

distribution in the jackrabbits. As I said, I should have known then that the lab was not the place for me, but it took me a while to come to this realization. It was not long, however, before I got lucky again and found myself at another field station.

My second field station experience came during the summer of 1965 when I was offered an opportunity to teach field parasitology at the W. K. Kellogg Biological Station (KBS) of Michigan State University. This facility is located about halfway between the cities of Kalamazoo and Battle Creek in south-western lower Michigan. Situated on the eastern shore of beautiful Gull Lake, the centerpiece of the Station is the elegant, Tudor-style summer home of W. K. Kellogg, the famous cereal-maker of Battle Creek (Fig. 1). My being hired was a huge piece of luck because it was eventually to lead my career in a new and completely different direction. As a National Institutes of Health postdoctoral fellow at the School of Public Health of the University of North Carolina in Chapel Hill, my office was next to that of Jim Hendricks. Jim was the most knowledgeable parasitologist I ever met and one of the nicest men I have had the good fortune of knowing. One day in the spring of 1965, Jim stopped at my office and asked if I needed a summer job. With a wife and two children to feed, I said yes, without hesitation, without thinking! Then, I asked, what kind of a job? He told me about the position at KBS, and that he knew the Director. The only problem was that the job had already been offered to Norman Levine and Martin Ulmer, both well-known, and well-established, parasitologists. I still thank my stars because they both turned down the job, and the Station's Director, George Lauff, hired me to teach field parasitology. He was to provide me with the chance of a lifetime, that of being associated with what was arguably the finest aquatic field station of its time.

My family and I were to return to KBS for 10 summers, where I was to teach field parasitology. When I began that summer of 1965, I knew nothing about the local fauna or of its parasites. In reality, I knew very little field parasitology. Despite having spent a summer at OUBS and shooting a lot of jackrabbits, my sundry field experiences were actually quite limited. I also knew very little about ecological parasitology. There were many things about this area of parasitology that were really quite new and intriguing. As I think about it now, before one can get started with this sort of teaching or research, a person must have a good idea about what comprises the local parasite fauna. In a letter to Miriam Rothschild, quoted in S. A. Ewing's (2001) *Wendell Krull: Trematodes and Naturalists*, Krull (a very good naturalist and field parasitologist) admonished that it was necessary "to know your territory with respect to its character and peculiarities. And the best way to learn the territory is to make a survey of the parasites of the animal or animals in which the investigator is interested. In this way one finds out which parasites predominate, the relative abundance of the flukes in the area and their distribution in animals in certain parts of the area." Although his advice was directed at those intending to do trematode life-cycle studies, it applies to anyone just beginning work in a new locality and searching for a problem on which to focus. So, that first summer, I was determined to find out what was there, although not with the idea of pursuing ecological parasitology, as that would come later. I was greatly helped by Dick Kocan and Joe Johnson, both of whom were knowledgeable about the area around Gull Lake. Dick was a grad student of David Clark at Michigan State University (MSU) and Joe was a field biologist who worked at KBS's bird sanctuary at nearby Wintergreen Lake. That first summer I began primarily with – what else? – gill netting and fish parasites. I was to learn a lot.

I quickly developed a passion for field parasitology, although I really did not know in my early years at Gull Lake what was beginning to take shape with regard to my career change. It was about this time, however, that I began to transfer concepts in the rapidly developing field of aquatic biology to understanding the nature of host–parasite relationships within an ecological context. Part of my learning came from actually doing work in the field, then of teaching in the laboratory. After all, if one is going to teach a new student about the parasite fauna in smallmouth bass, it is always best not to be surprised by what parasites the students may encounter. I was also to learn a great deal from the faculty at the Station. George Lauff had assembled as fine a group of aquatic ecologists as you could find in one place in the world, and that was not just my opinion. From these outstanding biologists, e.g., Don Hall, Earl Werner, Bob Wetzel, and Wayne Porter, among others, I learned a lot about ecological succession, primary production, cultural eutrophication, trophic dynamics, and predator–prey relationships. Whenever I could, I tried to take whatever they were saying and examine it in the context of host–parasite relationships.

The students during those summers at Gull Lake came from all over. Although many were from Michigan State University, there were a substantial number of high-school teachers who, under the auspices of a National Science Foundation summer program, were there to secure a Master's degree. These were among the best students I have taught in my 37 years in this business. They were very bright and always well-motivated: after all, they knew that with the Master's degree came a pretty good raise and, many times, opportunities for career advancement. There were even several Catholic nuns. The Sisters were really a lot of fun to be with, and one in particular, Sister James, I remember especially well. What a grand person! The students were supposed to do a special project, sort of a mini-thesis, as part of their degree requirement. Sister James decided she would work with me on some snail/trematode research I was doing at the time. She was quite a sight with her chest waders worn over the traditional habit and was frequently befriended by curious residents as she waded around the edge of Gull Lake searching for snails. As with the other nuns, she wore a rosary at her waist. Whenever she walked, the rosary would swing gently from side to side. I vividly recall the day it made contact with an electrical outlet on a lab table and exploded, showering beads all over the lab. As the beads were flying, I heard her shout, "Oh my God, I just killed Jesus!" She was rather embarrassed by this episode, but recovered. I think she even managed to get her rosary put back together.

I can also remember that the Station was a wonderful experience for our family. Our son, Craig, became very good friends with several of the entomology students those summers, as they would pay him a quarter for each butterfly he caught for their required collections. Those quarters bought a lot of candy bars. He actually became quite good at collecting and even rather knowledgeable about the critters. In the summer of 2000, my family and I toured the bowels of the British Museum of Natural History in London, thanks to a young American parasitologist friend, Pete Olsen, who was working at the Museum. A highly respected butterfly expert led part of our excursion and I was surprised when Craig began asking him questions about his work. They had a very good exchange about tiger swallowtails and other lepidopterans. There were other family times I recall at KBS, like one summer day when our daughter Lisa came running into our apartment exclaiming a fox was following her. My wife thought at first that she was playing a game, but quickly discovered that she was telling the truth when she saw the fox come trotting by the door. It had escaped from the Wintergreen Bird Sanctuary and had made its way over to the Station. Lisa was greatly disappointed when it was finally captured and returned to its cage at the Sanctuary. Our daughter fell in love with any sort of young animal she came across. She learned the hard way that not all of them were friendly. I recall her finding a young opossum, which she brought to our apartment to nurture because she thought its mother had abandoned it. She was to receive a rather vicious bite from the young creature, and thereafter she was rather careful about the animals she attempted to mother. Our youngest son, Charlie, learned to catch bluegills from the dock in front of the Station, and then how to "flay" (filet) them. He became quite a good fisherman, and still is, because of his summers at Gull Lake.

Even though I worked 65–70 hours per week each summer we were there, Saturdays and Sundays were for families. Ours swam a lot in the cold lake, or we might take in a movie at the drive-in theater over in Battle Creek. On Saturday nights, there was always a large crew from the Station that would drive over to Gilkey's Tavern just north of Hickory Corners, where we would eat one of their country dinners, down a few brews, and dance to the music of a three-piece band. The food was served American-style and the music was western. Both were always exceptionally good. Seminar nights were a treat for all of us. My wife, Ann, and I would hire a sitter for the children and she would always join me to listen to the exceptional group of speakers we had over those 10 years. There was usually a social event afterwards in the mess hall, an inexpensive way for husbands and wives to interact with the students and other colleagues. Even though research and teaching were the main focuses, there is something about a field station that pulls people together, be they students, faculty, or family. The camaraderie is quite real, powerful, and enduring.

Early in my Gull Lake career, I had some more luck when I developed what was to become a lifelong friendship and working relationship with an MSU graduate student by the name of J. Whitfield Gibbons. With a name like, that you knew he had to be from the south, say Alabama? More precisely, how about Tuscaloosa? Whit was an aspiring young herpetologist who was doing his doctoral research on the ecology of the painted turtle, *Chrysemys picta*. One of the things he was interested in was clutch size of the female turtles and the only way to learn about that then was to kill the turtle, open it up, and count the eggs in the uterus. It took a while, but I finally convinced Whit that the turtles also had parasites and why not give me the viscera of those animals he was required to kill so that I (we) could make use of the parasites. This work, and the publication that came from it in the *Journal of Parasitology,* started various sorts of collaboration that were to last off and on for more than 35 years, and continues to the present. It began with turtles, not unlike those Jim McDaniel and I had trapped 5 years earlier in a fish hatchery pond near Tischomingo, Oklahoma. What goes around comes around, and keeps on coming, as I was to learn later, especially with regard to turtles.

Something else happened during my KBS teaching days. Actually, I guess several things happened. I heard a seminar by one of the KBS faculty about the cultural eutrophication that was occurring in our own Gull Lake. It seems that the old septic systems around the lake were beginning to leak and this meant that phosphates were being added inadvertently, and at a rather rapid pace. I had also just read a eutrophication paper by Wincenty L. Wisniewski about parasites and parasitism in the vertebrates and invertebrates of a then (to me at least), obscure lake (Druzno) in Poland. I had no idea at the time that eutrophication and parasites would involve me with Wisniewski's study site in Poland nearly 30 years later. The significance of Lake Druzno will be the basis for my first parasite ecology field site discussion in Chapter 1. At any rate, on hearing the eutrophication seminar at KBS, it occurred to me that we had accumulated a substantial database on parasites in Gull Lake, which was undergoing eutrophication, and for nearby Wintergreen Lake which, according to Bob Wetzel (a very well-recognized limnologist), was already hypereutrophic. One thing led to another. On being encouraged by Robert P. (Mac) McIntosh (and, yes, a botanist) who was Editor of the *American Midland*

Naturalist, I published a paper in that journal comparing our system with Lake Druzno in Poland. (By the way, at the time of this writing Mac was still Editor of that journal, after 32 years in the job. This must be some sort of record for editors of scientific journals!)

Getting the eutrophication paper published was interesting too, since I had first submitted it to the *Journal of Parasitology*. Justus Mueller was the Editor at the time and the paper was rejected. In their collective wisdom, neither Justus nor his referees apparently understood it. Ecological parasitology was in its infancy and I honestly do not believe the referees or Justus made the connection between ecology and parasitology. That paper is still occasionally cited, more than 30 years later! I will always thank Mac and his referees for seeing the significance of the work. Counting the collaboration with Whit Gibbons on the turtle parasite study, it was the second publication in my new field of ecological parasitology.

Mac was eventually to help with the publication of another paper in the *American Midland Naturalist*, one that was ultimately to have considerable impact on the way in which parasite populations are categorized. Whit Gibbons and I had worked on the relationship between thermal stress in aquatic systems and parasitism down at the Savannah River Ecology Laboratory (SREL) for a couple of years after he finished at KBS. In 1970, we were invited by Elmer Noble to present some of our research in a symposium at the Second International Congress of Parasitology held in Washington, DC. Whereas this work was published in abstract form in 1971 as part of the Congress's proceedings, we felt it should be expanded and published as a theoretical paper. So, we approached Mac with the idea and he agreed at least to take a look at it when we were finished writing.

For nearly 3 years we labored with the manuscript, but could not get it right. The stress part was a "no-brainer." We had followed the line of Hans Selye and his ideas regarding what he termed the "general adaptation syndrome." In my mind, it was easy to jump from stress at the individual level to populations of free-living organisms, and even to ecosystems, using Selye's approach. But we were stuck on a point that had to do with the way in which parasitic helminth populations are organized. All of us would accept the idea that a population consists of a group of organisms of the same species occupying a given space and in which there is a free flow of genes. Moreover, most would agree that free-living populations may increase in size via both reproduction and immigration. Neither of these notions, however, has application to the populations of most parasitic helminths. Consider a small farm pond and the bluegill sunfishes within it, for example. We can easily visualize the bluegills as a population and understand that there should be gene flow among the bluegills. Then, consider the presence of adult nematodes, e.g., *Spinitectus carolini,* in the intestines of the bluegills in the pond. Do the *S. carolini* within a single host constitute a population, or do all of the *S. carolini* within the pond constitute a population? Which is it?

Whit and I agreed that it was neither. We realized that the numbers of *S. carolini* within an individual host cannot increase through reproduction, only by immigration (= recruitment). We also recognized that gene flow can occur among the *S. carolini* within an individual bluegill, but not between bluegills within the pond, at least not until another cycle of the parasite has been completed. Even then, the flow of genes is still restricted to those parasites within a single host. Free-living and many parasite populations are thus structured quite differently.

The recognition and resolution of this dilemma came during a conversation one summer evening with Whit in an empty cafeteria at KBS. During that long and intense discussion, the concepts of infra- and suprapopulations emerged. We designated the parasites within an individual host as an infrapopulation and all of the parasites of the same species in the pond as a suprapopulation. Once this idea was conceived, everything in our thinking about stress and parasitism seemed to fall in place. It was like we had broken a log jam. Although the ideas regarding stress and parasitism did not receive a great deal of attention, the parasite infra- and suprapopulation concepts that were developed have been widely accepted and are now routinely used in ecological parasitology.

In 1971, I went to Imperial College in London, not to do field parasitology, but to learn about the in vitro culture of cestodes from one of the great parasitologists of the time, or any other time for that matter, Professor J. Desmond Smyth. During our stay in London, I had a wonderful meeting with Clive Kennedy who had already established a reputation for himself as an ecological parasitologist working at Slapton Ley, on the coast of Devon. He was really good at what he did and my conversations with him were to lead me further down the road I was already, unknowingly, traveling. Clive and I had several long talks and, gradually, I realized I had become bored with biochemistry, which I had continued to do after leaving the University of Oklahoma.

Soon after returning to the USA in 1972, I read a paper by another Brit, H. D. (Harry) Crofton, of the University of Bristol. It was on the quantitative aspects of parasitism and was, as far as I am concerned, the seminal paper in modern parasite population ecology/epidemiology. I remember the summer when I first read it that I was totally fascinated with his approach to the manner in which he described parasites as being distributed in their host populations. I showed the paper to one of the ecologists at KBS, a really good one too, and his reaction was kind of, oh hum! In other words, he didn't get it. For a while, I wondered why, then it dawned on me that practically all organisms are distributed spatially in a nonrandom fashion, so what's new about parasites? or at least that is what I think he thought. What my friend did not see, and I had not bothered to explain to him, was that by concentrating the parasites in a few hosts, the risk of pathology, morbidity, and mortality within the host population, was spread unevenly. What I finally realized was that I had been pulled completely away from biochemistry to a whole new and different way of looking at host–parasite relationships, an ecological way. I can remember this moment of separation too, distinctly, when I made the conscious decision to switch my research entirely into the area of ecological parasitology, just like I had when I knew parasitology was going to be my life's work.

When I visited Clive in Exeter in the summer of 2000 to talk about his work at Slapton Ley, I asked him about Harry Crofton (who died in 1972 at a young age, and not long after his two important papers were published). Clive indicated that Crofton had done a lot of quantitative work on sheep nematodes, and had been struck by the fact that some possessed light worm burdens and some were heavily infected. This observation, apparently, led him into the overdispersion concept. Clive told me of their riding a train back from London, he to Exeter and Crofton to Bristol, after they had both attended a British Society for Parasitology Council meeting. As they visited, Clive recalled remarking about not being certain exactly what it was that constituted a parasite population. Crofton reflectively responded that he thought he did, and talked a bit about what was in the two (yet-to-be published) papers. He went on to tell Clive that he had been looking for a good data set with which to test his population ideas and that he had come across a paper by H. B. N. Hynes and W. L. Nickolas, which reported prevalence and abundance data for the duck acanthocephalan *Polymorphus minutus* in the amphipod *Gammarus pulex*. The two brilliant papers Crofton published in 1971 were wonderfully to illustrate his thinking about parasite populations and the dispersion concept. As Clive and I talked further, I was impressed when he emphasized that Crofton was not really an ecologist, yet his impact on parasite population ecology/epidemiology was enormous, both for that field of research, and for a gaggle of young (relatively, at least) parasitologists, including both Clive and me. As Clive said, "Crofton's quantitative approach was fundamental, because what he did was to show parasitologists why it was necessary to put numbers on life cycles."

After 10 marvelous summers at Gull Lake, my field experience was to shift south to SREL at the Savannah River Plant near Aiken, South Carolina. My herpetologist friend, Whit Gibbons, had finished his Ph.D. and taken up residence at SREL. This laboratory was started through the efforts of the leading ecologist of that era, the late Eugene P. Odum. The Savannah River Plant was the place where Odum had done his classic work on old field succession. But my interests were not in old fields. I was intrigued instead by Carolina Bays, cypress swamps, thermal pollution, and stress. So was Whit and, fortunately for us, so was the federal government, which funded the research at the Savannah River Plant (SRP) through what is now the Department of Energy, then the Energy Research and Development Administration, or ERDA. The SRP had to be, and still is, one of the most unique field sites in the world. There are roughly 300 square miles of land, completely surrounded by 10-foot chain-link fences on three sides and the Savannah River on the other. Access is limited and there are armed guards at the gates. Why? Because this is where weapons-grade plutonium is – or at least was then – produced by the US government. The locals referred to the place as the "bomb factory," which partly explains the security, e.g., the fences, guards, visible weapons, FBI clearances required to enter the site, and identification badges that had to be worn at all times. When my wife, Ann, and I went down to search for a house to buy in Aiken in the summer of 1974, we actually looked at several that had bomb shelters in their back yards! In those days, the Cold War was at its height, and the SRP was believed to be a likely target for nuking by the old Soviet Union.

One of the things that was really worthwhile about the SREL was being able to spend time with Whit Gibbons in the field. I can recall the first time I set foot on the site. Whit volunteered to take me and a couple of my new graduate students, Herman Eure and Joe Bourque, on a mini field trip. One of the first places we visited was the cypress swamp along the Savannah River. When we arrived at the swamp, Whit was into it immediately, like it was a wading pool in his backyard. My two students and I were very cautious as we edged out about 10 feet on a fallen pine tree over the murky black water, watching Whit closely as he waded in the swamp. After a couple of minutes, he asked us if we wanted to see a cottonmouth snake. We responded in the affirmative, and Whit said, "well, there's one there, and there, and there. . . ." It must have been humerous to watch three grown men trying to escape to solid ground in the middle of a swamp via a very narrow, fallen pine tree, and all at the same time. We were successful, but I am still absolutely amazed at how quickly we were able to move off that log without any of us getting wet, or getting any closer to a cottonmouth!

Another time, I remember driving on the plant site with Whit when he spied a coachwhip snake at the side of the road. Our truck screeched to a semi-halt and Whit was out the door, leaving the vehicle in gear, still moving, and us without a driver. Watching him catch the coachwhip was absolutely bizarre. The technique first involved approaching the snake from the rear. He grabbed the tail-end and slung the entire snake back through his legs. He then closed his legs and the length of the snake was pulled back through, quickly, and its head was caught as it passed between his legs from back to front. Obviously, the trick was not to get bit, anywhere! Whit was successful and brought his prize back to the truck, which, by that time, one of us had finally managed to stop. The snake was about 5 foot in length, but there was no place to put it for our 30-minute drive back to the lab (and no volunteers among us to hold it either). So, Whit found an old pair of waders in the truck and into one of the boots went the snake. I remember him handing me the boot and instructing me to squeeze the top of it tightly for our trip back to the lab, which I did, nervously (and he didn't need to tell me to keep a good hold on it). The only problem was, there was a hole in the boot, as we discovered on our return, and the snake was gone. The odd thing was, the snake could not be found in the truck either. We never did figure out what happened to it!

At SREL, my students and I began to exploit some of the field experience I had acquired at KBS as we delved into the ecology of acanthocephalans, cestodes, and trematodes in several different aquatic habitats. Herman Eure worked on aspects of the population biology of the tapeworm, *Proteocephalus ambloplitis,* and the acanthocephalan, *Neoechonorhynchus cylindratus,* in largemouth bass, with a focus on thermal pollution and thermal stress. He produced a couple of very good papers based on this research. Joe Bourque examined the population biology of several species of *Neoechinorhynchus* in the yellow-bellied slider *Trachemys* (= *Pseudemys*) *scripta,* and I was back to turtles again! Joe's research was also published and well received. Other students of mine, Joe Camp,

John Aho, Jim Coggins, and Kym Jacobson, did all, or at least some, of their M.S. or Ph.D. work down at SREL, and all four were quite successful in their efforts.

My own research at SREL was also eventually to focus in Par Pond – not really a pond, but a very large cooling reservoir with a huge population of very catchable largemouth bass. For most of this work, we preferred to electrofish over the traditional method of bait fishing so that we could obtain an adequate sample size for the work we were doing. At that time, one of my graduate students, Terry Hazen, and I were attempting to understand the epizootiology of red sore disease in fish, caused by *Aeromonas hydrophila*, a Gram-negative bacterium. Normally more or less ubiquitous in aquatic habitats and usually not a problem unless the fish become stressed, *A. hydrophila* can be lethal and cause massive fish kills. I had met Terry, also known as Big T (among other things), while he was a graduate student at KBS. He always referred to me as "god" (among other things) and my wife, Ann, as Mrs. "god." He accompanied me to SREL in 1974–75, and then did his Ph.D. with me on the ecology of *A. hydrophila* up in North Carolina. As part of our study at SREL, however, Terry and I captured, marked, and released close to 2000 largemouth bass in the Par Pond reservoir in one year. Terry would drive the boat and I would man the nets and electricity on a platform at the front end.

Electrofishing in the reservoir was not only fun but, on occasion, exciting as well. I do not think I have ever been so startled in the field as I was when we crossed paths with an 8-foot alligator! It was not too happy when I hit it with the electricity, but the 'gator disappeared immediately when I cut off the voltage.

Fishing in Par Pond by rod and reel was not very sporting because the bass were so easy to catch. But, we still managed to get in some angling. The very first time I went out on Par Pond, I remember getting my line tangled and, while trying to get it fixed, caught a 3-pound bass as my hook – a baitless hook at that – dangled in the water over the side of our boat! One evening, I caught 37 largemouth bass, all greater than 2 pounds, in less than 45 minutes. My arms were sore the next day from hauling in so many fish in such a short period of time. I know these stories sound like tall tales from an old fisherman, but I swear they are true.

Whereas our work was quite rewarding, and successful, perhaps the most fascinating thing to come out of my personal experience at SREL was to witness the ingenious research of Whit Gibbons in the Carolina bays. A Carolina bay is a curious geological feature of the coastal plains in North Carolina, South Carolina, and Georgia. These ephemeral ponds are shallow depressions, rather wide at one end and narrow at the other, always oriented in a north-west–south-east direction. Some people have speculated they were created by a meteor shower long ago. Whatever their origin, during the rainy season they fill with water and, of course, herps of all kinds. During extended drought periods most of them dry up. Their surface areas range in size from several hundred square meters up to a square mile. I can recall one day in 1975 when Whit decided he was going to construct a drift fence around one of these Carolina bays. A drift fence is a rather simple structure made from sheets of aluminum that are placed in the ground in such a way as to prevent any animal from moving into, or out of, the bay. At intervals of about 10 meters, five-gallon buckets are sunk into the ground. Any migrating animal on encountering the fence will turn and move lateral with it, eventually come to a bucket and fall in. All one has to do then is walk the fence each day, remove any critter that has fallen into the bucket, record its identity, and release it on the other side of the fence.

Whit was primarily interested in turtles of all kinds, but, of course, during the spring rains, enormous numbers of different kinds of herps, thousands of them (mostly salamanders), are on the move. Based on this research, a number of important discoveries were made regarding the ecology and population biology of the slider turtles and other herptiles present in the Carolina bays. One the most innovative techniques used by Whit was his method of measuring clutch size in female sliders, a very important parameter in any population study. Up at KBS when he worked with painted turtles, he was required to kill a certain number of his females in order to measure clutch size, which, as I explained earlier, is why he and I first began collaborating on research. Down at SREL, he discovered that to measure clutch size in a female slider, all he needed to do was X-ray it and then count the eggs in utero on the X-ray film, without destroying the animal – very clever indeed. He also learned that he could Xerox the plastron of each turtle he caught, giving him a virtual fingerprint of each individual to assist in later identification when it was recaptured, again, and again. Whit, or one of his technicians or students, ran the drift fence around one particular Carolina bay for some 13 years, marking all the turtles as they moved in and out, and recapturing many in the process. Using this mark–recapture technique, he was able to show, among many other things, that the life span of the yellow-bellied slider turtle was far longer than anyone had previously suspected, ranging up to

Figure 2   A group of students on the shore of Belews Lake, Stokes County, North Carolina. This large reservoir was subjected to selenium pollution, which permitted several of my graduate students the opportunity of examining the parasite population dynamics and copepod community structure under unusual conditions. (Photographer unknown.)

40 years! These studies are undoubtedly among the most thorough on any reptile of which I am aware.

After all the field research we had done in Oklahoma, Michigan, and South Carolina, my students and I finally began to do some work in North Carolina. In about 1970, the Duke Power Company began construction of a cooling reservoir, called Belews Lake (Fig. 2), about 20 miles northeast of Winston-Salem. As Chairman of the Biology Department at Wake Forest University in 1975, I figured we could do some field work in this new lake if we could persuade Duke Power to give us some acreage on which to construct some field facilities of our own, which they did. We named it the Charles M. Allen Field Station, in honor of one of our departmental colleagues who was so instrumental in helping us get started out there. The lake is large, some 1580 ha in surface area, 180 feet deep in a couple of spots, and exceptionally beautiful. Duke Power retained the property rights around the lake, thereby minimizing nonpoint source run-off from farming operations. This, combined with its very small drainage basin, meant that the lake was never very productive. Even so, by 1976, there was something like 27 species of fishes in the reservoir – a fairly typical piscine community for a large reservoir in the south-eastern part of North America. However, much to our chagrin, within just a couple of years, the number of fish species fell to just two. Mosquitofish and fathead minnows, both of which are highly tolerant of many toxic agents, were all that remained. Our beautiful lake had become polluted.

Duke Power burn coal from their own West Virginia mines to create steam to run the turbines in their coal-fired operation. The fly ash left after burning the coal in those early days was made into a slurry and then pumped into a settling basin nearby. After the particulates had settled out, the water was allowed to return to the lake. The pollution in Belews Lake was created by high levels of selenium in coal burned to heat the boilers at the plant. A required trace element in most biological systems, selenium, when present in high concentrations, will become exceedingly toxic to vertebrate animals. Like mercury and DDT, it will be biomagnified as it moves through the typical food chain. Present in high concentrations in the fly ash, the selenium was dissolved in the water, which was coming back into the lake from the settling basin. Duke Power did not realize what they were doing and, within 2 years, the fish community in the main body of the reservoir was completely decimated.

Bill Granath began working in Belews Lake after the fish community had been reduced to virtually nothing. Prior to the lake's biological demise, we knew there were at least six species of helminth parasites present. With the elimination of most of the fish, I assumed the parasites had also been eliminated and insisted there were not any parasites left. Bill was (is) a fine, but rather stubborn soul, and was determined on going out and having a look anyway. Naturally, I was completely flabbergasted when he came into my office one day and reported that he had found a tapeworm in some of the mosquitofish he had necropsied. There should not have been any parasites present in the lake and certainly no cestodes in mosquitofish. He eventually identified them as *Bothriocephalus acheilognathi*, the Asian tapeworm, which had been described by Yamaguti in Japan some 50 years before and, I would add, a parasite that had not been previously present in Belews Lake. This cestode is what one might call the ultimate opportunist among fish parasites. First described in Japan, it spread initially into Eastern Europe, then to the UK and North America. Its movement was presumably in connection with the huge, worldwide trade in exotic fishes, a largely unregulated practice with potentially devastating effects, depending on the host (and parasite) being moved from one continent to another. Glenn Hoffman, a North American fish-parasite expert, opined that *B. acheilognathi* was introduced into North America sometime in the mid 1970s in imported grass carp. It then became established in the fish bait-farming region of Arkansas. How did it get from Arkansas to Belews Lake in North Carolina? Our best guess was that a bass fisherman bought some Arkansas red shiners at a local bait shop and used them at the lake. Since he had obviously had no luck in catching any large-mouth bass, he dumped the red shiners into the lake in frustration and, with them, the Asian tapeworm.

I mentioned that this cestode is the ultimate opportunist. It has been reported to mature sexually in more than 40 species of fish hosts and to use at least five different species of copepod as its intermediate host, which explains why it was able to establish once it was introduced into our lake. The parasite is also interesting since, in small fish hosts, it can be lethal due to the explosive growth of small enteric plerocercoids once stimulated by rising temperatures each spring. Smaller fishes cannot support a large number of adult cestodes – the gut becomes occluded, and the fishes die as a result.

The first research on the Asian tapeworm in Belews Lake was done by my Ph.D. student, Bill Granath, who was interested in the effects of thermal pollution on the population biology of the parasite. It was an exceptionally solid effort and was followed by some more very good work by two other graduate students, Dave Marcogliese and Mike Riggs. Dave's attention was mainly on the population and community ecology of the copepod intermediate host, whereas Mike was to examine another facet of the parasite's population ecology. In pursuing their dissertation research, each of them was to take advantage of an unusual characteristic of Belews Lake. I said before that the lake's fish community declined from 27 species to just two in the span of just a couple of years. This was not exactly true. Two small streams feed the lake, with Belews Creek being the main source of water. There are two arms that join, forming the main body of the lake. One of these arms is thermally altered by the power-generating operations. Water from the heated arm runs back into the main body of the lake through a man-made ditch, or canal. There, the water cools, and is then cycled back through the power-generating plant.

About 3 miles away from the canal, Mike Riggs discovered the site of a most unusual physical phenomenon in the lake. There is a railroad bridge that crosses a narrow part of the main arm. When Mike went about 50 m north of the bridge and dropped a Sechi disk into the water, there was a reading of about 1.5 m. For most reservoirs in the south-eastern USA, this is relatively high, reflecting the very low turbidity in the main body of Belews Lake. However, when Mike went about 50 m south of the bridge and took a reading with the Sechi disk, the measurement was about 50 cm. This is much more in line with other south-eastern reservoirs, which are generally rather turbid bodies of water. In other words, north of the bridge, the water color was a gorgeous blue-green and relatively clear. South of the bridge, however, it was chocolate brown, and very muddy in appearance. It was almost like there were two lakes. Indeed, there were, and still are, not only because there was a huge difference in turbidity and water quality, but because the headwaters arm of the lake south of the bridge possessed a full community of fishes, exactly like the one which was wiped out by the selenium poisoning in the main body of the lake. Mike determined that the separation between the contiguous bodies of water was created by a chemocline, a stratification gradient that resembles a thermocline, except that a chemocline is due to a chemical gradient and a thermocline to a temperature gradient. This chemocline was powerful enough to prevent the selenium from extending into the lake's headwaters and from eliminating the fish community there as a result. Since the poisoned part of the lake had no piscine predators, Mike was in a position to look at the population biology of the tapeworm in a host without the effects that might be created by piscine predators.

Dave's research was focused on the community ecology of the copepod intermediate hosts of the Asian tapeworm. Ron Dimock, a colleague of mine in the Biology Department at Wake Forest, and I had made extensive plankton collections in Belews Lake in 1975–76 as part of another study, but had done nothing with the samples, preserving all of them in 70% ethanol. After the selenium pollution, the frustrated bass fisherman introduced red shiners. In the absence of piscine predators, the red shiner population exploded, driving the mosquitofish and fathead minnow populations down to the levels they were during the pre-selenium days of the lake. The planktivorous red shiners became dominant and, as a result, altered the copepod community structure dramatically. So, Dave Marcogliese had an opportunity to look at the effects that heavy predation had wrought on the plankton community in Belews Lake.

It's interesting to look back on these Belews Lake studies because they show that research can sometimes become almost like a "stream of consciousness." In this sense, Bill was not looking for the Asian tapeworm when he went to Belews Lake the first time and seined some tiny mosquitofish for necropsy, but he found the parasite and exploited its presence. When Ron Dimock and I collected the plankton in 1971–72, we did not anticipate that the lake would become polluted with selenium, that the fish community would be so radically altered, or that the planktivorous red shiner would become the dominant fish species for a period of time in the main body of the lake. All of these events allowed Dave to do his thing. Similarly, there was no way of predicting the development of the peculiar chemocline which encouraged Mike to follow the parasite's population dynamics both in a complex community with 27 species of fish and in a predator-free system dominated by the planktivorous red shiner. I think this style of doing research, the "stream of consciousness" approach, is probably the most fun of any that can be accomplished because it allows you the total freedom to go wherever the research takes you. A person can be totally unrestrained, almost by design!

Even though we used this approach in Belews Lake, the best example of this sort of research is based on work we were subsequently to undertake in a small farm pond located within just a few meters of Belews Lake. Back in 1983, a young undergraduate student, Amy Crews, who had been accepted into our Master's program at Wake Forest, approached me and said she wanted to get an early start on her thesis research. What could she do? I responded by saying that I knew about this small farm pond out near Belews Lake, and perhaps we could collect some snails and look to see what cercariae they were shedding. Maybe this would give her a start. Little did we know where this short field trip would take her over the next 2 years, or some 15 other graduate students who subsequently worked there over the next 19 years. Amy and I went out to this pond and collected some *Helisoma anceps*, a common pulmonate snail in our part of the country. We brought them back to the lab and isolated them in plastic jars, then waited 24 hours and examined them for shedding cercariae, the usual procedure. She came to my office the next day and reported that there were some swimming cercariae, but that there were also some lifeless (in appearance at least) objects at the bottom of several jars and would I come to the lab and take a look, which I did. I had never seen anything like them before, but we did some investigating and found them to be the cystophorous cercariae of a hemiurid fluke named *Halipegus occidualis*. Amy's preliminary observations indicated that close to 60% of the snails were infected with this parasite and that all of them with patent infections were totally castrated. Her Master's research was laid out for her after our short trip to the small pond. It also opened the door for what I consider to be one of the most extraordinary opportunities anyone working with the ecology of host–parasite relationships could possibly have. I think it is an interesting story, one which will become the topic of another essay to follow.

All of this fieldwork led me to write the present book. In each of the following chapters, I will talk some science, tell some stories about some of the folks who did the work, and get into some personal philosophy about my area of parasitology and biology. Another reason for undertaking this project was to describe several of the most famous field sites where ecological parasitology has been done. Please don't get me wrong from my selection of these particular sites. Field parasitology has actually been done in many places and for a long period of time. Indeed, the early Egyptians described what we now believe were malarial fevers, *Ascaris lumbricoides*, and pink urine caused by *Schistosoma haematobium*. These are all epidemiological in character and deal with human parasites. They are none the less ecology. Moreover, virtually every parasite survey, whether of fishes or frogs, has an ecological connotation of some sort. My selection of field sites in this book is based entirely on my personal bias. I have visited most of the sites, with Lake Druzno in Poland and Algonquin Park and some other places in Canada being exceptions. Among the other locations I have written about are Slapton Ley on the Devon coast of England, Douglas Lake in Michigan, Cedar Point Biological Station in Keith County, Nebraska, and Charlie's Pond in North Carolina. There are well-known field stations at a couple of these places. Others are field sites where some major work was conducted over the long term, or where some specific study was done which led to a long-lasting impact on ecological parasitology. Obviously, there are both stations and localities where other major research has been carried out and a few of these will be mentioned where appropriate, but they will not be in the main theme. As I said in the Preface, most of what I write about has come from my own personal experience.

People with a real focus conducted a great deal of interesting work at these sites. So, I wanted to look at what drew these investigators to these sites. Was it a common experience that led them to these locations? Was it accidental, a matter of convenience, or a stream of consciousness, like that which prompted our work in Belews Lake or Charlie's Pond? In the case of the two field stations, about which I will write in later chapters, what made them so attractive as field parasitology sites? In fact, one of them, UMBS at Douglas Lake, became the mecca for nearly every major American parasitologist of the early twentieth century. Keith County is home to the Cedar Point Biological Station of the University of Nebraska where John Janovy, Jr. and Brent Nickol, two old friends, have done so much work out on the high plains of western North America. I also wanted to include something with an epidemiological focus. Two years ago, I visited Ana Flisser in Mexico City and the National University of Mexico where I met with Carlos Larralde and Edda Sciutto (a husband- and-wife team who were to become my good friends). They took me into some fascinating and interesting areas south of Mexico City where neurocysticercosis is endemic. I will write about "the mystery disease of Pudoc," based on some input from Darwin Murrell. Peter Hotez provided me with several interesting stories about his work on hookworm disease in China.

Finally, I must ask, rhetorically, why would anyone want to sit down at a PC and take a huge amount of time over a period of more than 2 years to write a book, especially one like this? It is not a scientific tome, although it certainly has a lot of science in it. It is not really biographical or autobiographical, although there is some of both in it. It is not a historical account of field parasitology, even though there is clearly a historical component to it. Although I had toyed with the idea for this book for several years, the real momentum for writing came when I began thinking about some of those parasitologists of the twentieth century who influenced my career. I realized that many of them were "dying off," or were nearing retirement, and would soon be leaving the so-called "academy," one way or another. I felt that many of these people, because of their impact on my career, might have some interesting stories to tell about their work, stories that might influence a student, or help someone develop some important insights with regard to our science. In retrospect, I wish, for example, that I had spent some time with Doc Stabler and Professor Self (Fig. 3), just to sit with them and talk about their lives and experiences. But I did not do it. With all of this in mind, I began to write. Then, in August of 2001, I learned that the highly respected Canadian parasitologist, Roy C. Anderson, had died, unexpectedly. In fact, his fatal heart attack came less than a day after I had e-mailed him with a request that he share some stories about his work at the Wildlife Research Station in Algonquin Park. It was then, the moment when I learned of Roy's death from Sherwin Desser, that I knew I had made the right decision, to write this book. There was some really good parasitology done in the last century, by some really good parasitologists, and a lot of it was accomplished in the field. I hope that the contents of this small book will contribute to a better understanding of all this fieldwork, and provide some insight into the lives of a few of those who did it.

Figure 3   Profesor J. Teague Self, in his office in the Department of Zoology, University of Oklahoma, Norman, Oklahoma. This giant of a man directed me through the rigors of obtaining the M.S. and Ph.D. degrees from 1958 to 1963. I am forever grateful to him for all of his help and insight. (Photographer unknown.)

References

Ewing, S. A. (2001). *Wendell Krull: Trematodes and Naturalists*. Stillwater, Oklahoma: Oklahoma State University.
Holloway, M. (1996). Profile: Miriam Rothschild. *Scientific American*, May, 36–8.
Wardle, R. A. and McLeod, J. A. (1952). *The Zoology of Tapeworms*. Minneapolis, Minnesota: The University of Minnesota Press.
Yamaguti, S. (1958). *Systema Helminthum*, vol. 1. New York, New York: Interscience.

Suggested readings

Crofton, H. D. (1971). A quantitative approach to parasitism. *Parasitology*, **62**, 179–93.
Crofton, H. D. (1971). A model for host–parasite relationships. *Parasitology*, **63**, 343–64.
Esch, G. W., Gibbons, J. W., and Bourque, J. E. (1975). An analysis of the relationship between stress and parasitism. *American Midland Naturalist*, **93**, 539–53.
Gibbons, J. W. (1990). *Life History and Ecology of the Slider Turtle*. Washington, DC: Smithsonian Institution Press.

1

From Elbing in Kansas, to Elblag in Poland

> The great end of life is not knowledge but action.
>
> THOMAS HENRY HUXLEY, *TECHNICAL EDUCATION* (1877)

The Louisiana Purchase from Napoleon in 1803 was a bold move on the part of Thomas Jefferson, a natural beginning of what was to become the doctrine of Manifest Destiny for the (then fledgling) USA. He sent Meriwether Lewis and William Clark to explore it, collect and describe the plants and animals, and try to persuade the native Americans that they had nothing to fear from those who were sure to follow. The book, *Undaunted Courage*, by the late Steven Ambrose (1996), on the adventures of these two great naturalists, was absolutely fascinating, especially his re-counting of the descriptions of the flora and fauna these two explorers encountered in their travels across the western half of the North American continent.

After the Lewis and Clark expedition, opening the trans-Mississippi west began almost immediately, with the incursion of a small, but ex-ceedingly intrepid, group of fur trappers and explorers in the first 30–35 years of the nineteenth century. These so-called Mountain Men, like John Colter, Jedediah Smith, Jim Bridger, and William Sublette, trapped beaver, lived with and fought the American Indians, and traveled exten-sively throughout the western wilderness. Word of their adventures, and tales of what these trappers found and saw, spread east, creating a near sense of urgency for other adventurers to follow and see these new ter-ritories for themselves. Some of what those in the east heard from these Mountain Men surely must have seemed like outrageous exaggerations. And, having read several of their diaries myself, I agree that they did have a tendency to manufacture some tall tales on occasion. After all, who in their right mind would believe that there was a place where huge plumes of superheated water and steam were vented from the earth nearly 100 foot into the air, almost on the hour, every hour of the day and night, in all seasons of the year? Or, what about those snow-capped mountains that were so high they seemed to touch the sky? Then, who could imagine so many buffalo that a person standing high on a hill could not see all the way across a single herd, or that a migrating herd could be so large it would take several days before it passed by completely?

About 15 years after the days of the Mountain Men were over, the west really opened up and great numbers of people headed in that direction, for, in 1849, gold was discovered at Sutter's Mill in California. Hundreds of thousands of people were then on the move. Some took fast clipper ships around Tierra del Fuego, the southern tip of South America, and up to California. Others were to sail down to, and cross, the Isthmus of Panama, then sail up the western coast of Central America and Mexico. A decade later, this was followed by gold and silver strikes in Colorado, and the slogan "Pike's Peak or Bust" was to emerge as a kind of summons for all sorts of new people from the east. Both times, however, teams of oxen or mules pulled sturdy Conestogas in great wagon trains over the Sante Fe or Oregon Trails, many times guided by the same Mountain Men who had opened the west. Even as these migrations crossed the Great Plains, most of the pioneers had not the slightest inkling they were on some of the most fertile farming land on the planet. Indeed, General Zebulon Pike, for whom Pike's Peak was named and one of the great explorers of the early nineteenth century, described the eastern part of what is now the state of Colorado as the "Great American Desert."

At last, however, came the construction of railroads across the vast expanse of the western USA in the late 1860s, following the bloody Civil War, and there was a massive influx to settle the farmlands of the Great Plains. Many in this latter group were immigrants from eastern Europe. Among them were those who brought a special type of wheat from Russia that made Kansas and the other plains states into what became known as the "bread-basket" of the world. The primary lure for this last explosive movement of people was provided by a combination of the railroads and the federal government. Thus, there was the promise of 160 acres of land for each family if they agreed to build a home, often of sod, and then farm it for 5 years, at which time they would receive title to the property. The idea was that these homesteaders would grow wheat and other grain crops, and raise cattle and sheep, all of which would then be shipped back east on the railroads to help feed the rapidly expanding population. The country had indeed benefited from Jefferson's Louisiana Purchase and from the theme of Manifest Destiny, well, almost everyone. The native Americans paid the greatest price. They were gradually driven off their lands, usually through fraud, or because of broken treaties by the American government, and ultimately settled on prescribed reservations, many of them far less than one-hundredth the size of the territory they originally roamed. But this is yet another story, and a very sad one at that.

Many of the immigrants who came west in those years had fled their European homes because of religious persecution. Among these refugees was a contingent of Mennonites (Anabaptists) from what is now Poland. At the time, however, that area of eastern Europe was part of East Prussia since, for a period of about 300 years, Poland had ceased to exist as a sovereign nation. These German-speaking people had originally migrated from Belgium to that part of eastern Europe in the sixteenth century to escape the extremes of the Spanish Inquisition. Between 1876 and 1880, several families of Mennonites homesteaded on some of the rich farmland in Butler County, near the center of the state of Kansas. In 1885, the Rock Island Railroad decided to build a rail line that was to extend from the tiny community of Peabody, Kansas, down to Wichita, and then on to the Gulf coast of Texas. In constructing the railroad, the Rock Island also decided to build a series of depots to make it easier to ship goods into the rural areas and transport farm products out. The first of these depots was to be located about 7 miles south of Peabody. The land for the depot was acquired from Jacob W. Regier, one of the Mennonites who had migrated from East Prussia. Then, with construction of the depot, the local farmers decided they wanted an organized community, so Regier sold 80 acres of his land for $5000 to a newly formed town company. Naturally, they needed a name for the town and someone suggested it be in honor of Regier, but old Jacob rejected the idea. Instead, he suggested the town be called Danzig, Elbing, or Marienburg, all cities in which Regier had lived while in East Prussia. The name of Elbing was finally assigned to the tiny community because the English-speaking railroaders said it was the easiest of the three choices to spell.

I have been somewhat acquainted with Elbing in Kansas all of my life, having spent my youth only 25 miles away in Wichita. Later, when I married Ann Speir, I was to learn that one set of her great-grandparents, Michael and Sophronia Sophia Guinty had immigrated from Ireland in the 1840s and had settled in Elbing. They, along with the Mennonites, are

Figure 4 Pleasant View Cemetery in Elbing, Butler County, Kansas, where the Mennonites from Elblag (at the northern end of Lake Druzno), Poland are buried, along with the Guintys, my wife's maternal grandparents, and immigrants from Ireland.

all now buried in the Elbing's Pleasant View Cemetery (Fig. 4). Although I had known about Elbing all of my life, I must admit that I actually knew very little about the small Kansas farm town until I began to read extensively about Poland in the spring of 2000. It was then I learned that Elbing in what was East Prussia was now called Elblag and is currently part of Poland. In the same way, Danzig is Gdansk, and Marienburg is now Malbork. Gdansk is, of course, the place where Lech Walesa led the uprisings by the workers in the Lenin shipyards that eventually brought the downfall of communism in Poland. Malbork is the site of a famous castle, built and controlled by the Teutonic Knights who maintained a strong grip on the north-western and north central parts of what is now Poland for several hundred years, beginning in the thirteenth century.

I have explained my personal relationship to the community of Elbing, Kansas, but what of its namesake, Elblag, in Poland? This city is of interest to our story regarding famous parasitology field sites because it sits at the north end of Lake Druzno, which is where Wincenty Wisniewski undertook the first investigation on the relationship between eutrophication and parasitism, an ecological study that, in my humble opinion, was far ahead of its time. Like many small farming communities of the

Great Plains in the USA, not much is left of Elbing, Kansas, except the old cemetery. In contrast to Elbing, Elblag in Poland was settled in 1297, and is presently a city of about 130 000 residents. It is situated in north central Poland, an area called the Vistula Delta. Early in its history, it was an important city under control of the Teutonic Knights. As a port on the Baltic Sea and a member of the powerful Hanseatic League in the fourteenth and fifteenth centuries, it was, and still is, connected to the Baltic via the Vistula Lagoon. The city was overrun by the Germans at the start of World War II and became part of Hitler's Nazi Germany. Elblag (Elbing) was the main site of German U-boat production and was almost totally destroyed by allied bombing and during its capture by the Soviets toward the end of the war. The Prussian city of Elbing became the Polish city of Elblag with the restoration of the state of Poland after World War II.

Lake Druzno sits just to the south of Elblag. This body of water forms the northern terminus of the 82-km-long Elblag–Ostroda Canal. The canal was constructed in 1848–60 under the direction of Georg Steenke, a Königsberg engineer who also designed it and convinced the King of Prussia that it should be built. It is a real engineering marvel and the longest canal still operating in Poland. At the southern end of the canal is the city of Ostroda. Since medieval times, Ostroda has been surrounded by magnificent forests, much in demand as timber by the merchants of Gdansk and Elblag. Prior to construction of the canal, the only way in which cut timber from the forests near Ostroda could be moved to the more northern cities was by a lengthy water route through Torun via the Vistula and Drweca Rivers. Most believed that a canal using a system of conventional locks between Ostroda and Elblag could not be built because of the rugged terrain. Steenke thought otherwise. His design for the canal was to take advantage of the six lakes located between the two cities. Near Elblag and Lake Druzno, he created five slipways, each of which uses two trolleys to carry boats and their cargoes over the dry land separating the waterways. The canal was badly damaged in World War II, but within a year of the end of the war it was back in operation, carrying both timber and tourists.

Wincenty Wisniewski was the driving force for the early research on the relationship between parasitism and eutrophication in Lake Druzno. This interesting man was born on 14 September 1904 in Dobromil, near Lwow, in what was to become eastern Poland immediately after World War I. At Potsdam following World War II, Stalin insisted on acquiring vast areas of what had at one time been eastern Poland, including Dobromil and Lwow, and incorporated them into what are now the Ukraine and Belarus. Simultaneously, a large section of what had been part of Germany in eastern Prussia, prior to the war, was transferred to Poland, including the areas around Elblag and Lake Druzno.

After attending and graduating from schools in Lwow, Wisniewski traveled to Warsaw, where he began undergraduate work at Warsaw University. There, he was to come under the influence of the great Polish parasitologist, Professor K. Janicki, who was instrumental in working out the life cycle of *Diphyllobothrium latum*, the so-called broad fish tapeworm. Wisniewski completed his undergraduate work in 1927 and then began a remarkable teaching and research career, first serving on the Agriculture Faculty at the Engineering College in Lwow. In 1930, he received his Ph.D. at Warsaw University and, by 1937, he had become an assistant professor there. As were most of the other able-bodied Polish males, he was mobilized into their army in August of 1939. On 1 September 1939, Hitler struck at Poland, brutally and without provocation, as part of his strategy to conquer and occupy the whole of the European continent. Although the Poles fought hard and with great courage, they were overwhelmed quickly and savagely by the superior Nazi forces. Wisniewski was with the 79th regiment in General Kleberg's army. On 5 October 1939, he was captured by the Nazis at Kock, north of Lublin, where, as he puts it, he was "taken into slavery." In a brief autobiography obtained from his daughter Katarzyna, he indicates that he managed to escape from the Germans within a few days of his capture (although he does not say how he did it). He headed for Warsaw, but that city had been overrun and by that time was under the complete control of the Nazis. It was then he decided to join the Free Polish Army that was being formed in Palestine. He traveled south toward Czechoslovakia, but had the misfortune of being captured by Soviet border guards in the Karpat Mountains. As was the case of many Poles at the time, the Russians sentenced him to serve for 3 years in a Siberian labor camp at Charkov, and off he went. As it turned out, he was immensely fortunate to receive the prison sentence since Stalin was subsequently to murder some 20 000 army officers and intelligentsia at Katyn after he and Hitler annexed Poland in 1939.

Of course, Hitler then launched Operation Barbarosa and attacked the Soviet Union on 22 June 1941. The expedient Stalin needed all the help he could get in his brawl with Hitler, so he and the Polish government in exile signed a mutual assistance treaty in July of 1941. General Wladyslaw

Anders agitated Stalin until he freed some 160 000 Polish men, women, and children, among them Wincenty Wisniewski, who made their way down through Persia (Iran) and into what was then Palestine. Wisniewski joined with the Polish Corps of General Anders, which fought with great valor alongside the British Eighth Army at Tobruk in North Africa and later in Italy at Casino. At the war's end, Wisniewski found himself in England where he was eventually demobilized in 1947. He returned to Poland immediately and resumed his position as Assistant Professor of Zoology at Warsaw University in May. Within 3 months of his return, he was made Professor and Head of the General Zoology Cathedral at Warsaw University. It took him just 3 months to go from Assistant Professor to Professor, because so many of the Polish academics had been murdered by Hitler or Stalin, or had been killed fighting the Germans in North Africa or Italy. During his highly distinguished career, Wisniewski was to hold a number of teaching, research, and administrative positions in various Polish universities and the Polish Academy of Sciences.

Most of Wisniewski's work prior to World War II was focused on the biology of trematode cercariae. But, as he said in his brief autobiography, "tramping around the world" during World War II extended his interests to "ecological problems." He began a series of investigations on the biocoenoses of lakes, with an emphasis on parasites. When Wisniewski refers to a biocoenosis, he is talking about the nature of the parasite species mix within a habitat, in this instance Lake Druzno. The idea of parasites (or a parasitocoenosis) within the framework of a biocoenosis had its origins with the Russian parasitologist, E. N. Pavlovski, who believed that parasites, together with their hosts, comprise a special portion of the biocoenosis and, therefore, a part of their biotope. His idea was that parasites are linked to the overall biocoenosis via their hosts and that their circulation in the biotope is mediated through the food chains. Tied directly to this concept is the notion of landscape epidemiology. Wisniewski attempted to examine the parasite fauna in Lake Druzno (Fig. 5) and link it to eutrophication within the biocoenosis context. This effort was to prompt a series of studies by several other investigators over the next 25 years or so, primarily in the UK, Scandanavia, and North America. All of these investigations in various ways also addressed the relationship of the trophic status of lakes and the nature of their parasite faunas. Because of Wisniewski's far-sighted effort in this area of ecological parasitology, I decided to begin this collection of essays with a discussion of his work and some of those who followed his lead.

Figure 5   Lake Druzno, Poland, just south of Elblag. This lake was the site of some famous field studies conducted by Wincenty Wisniewski by in the early 1950s. (Courtesy of Jerzy Rockicki.)

Prior to dealing with the specifics of eutrophication and parasitism, however, and the contributions made by Wisniewski, it is necessary that we first examine the nature of eutrophication itself. Eutrophy is derived from a German word, *eutrophe*, which, quite literally, means nutrient-rich. The opposite of eutrophy is oligotrophy, which refers to something that is nutrient-poor. When reference is made to nutrients in aquatic systems, simultaneous consideration is usually given to primary production and carbon fixation via photosynthesis. In aquatic ecosystems, plants require a range of nutrients for primary production, with the three most dominant being carbon, nitrogen, and phosphorus. In most bodies of water, carbon and nitrogen are abundant, and not in short supply. The limiting nutrient in most aquatic systems is usually phosphorus, which is essential in the synthesis of both adenosine triphosphate (ATP) and nucleic acids. Eutrophication, therefore, is associated with nutrient (phosphorus) enrichment, mostly from allocthonous, or outside, sources. In more general terms, eutrophication is a form of ecological succession. When ecological succession occurs, it is a directional process, involving predictable change in community structure and dynamics. Although it is a natural phenomenon, eutrophication can be speeded up in aquatic systems through

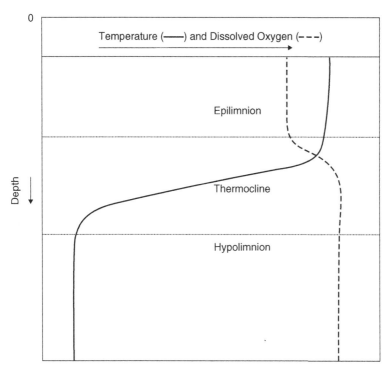

Figure 6 Thermal and dissolved oxygen profiles of an ideal north-temperate oligotrophic lake in mid-summer. (Courtesy of Joel Fellis.)

the addition of phosphorus via the activities of humans, in which case we have cultural eutrophication.

To understand all of the ramifications of eutrophication, it is first necessary to describe some of the physical attributes of an oligotrophic, or nutrient-poor, aquatic ecosystem since succession normally proceeds from oligotrophy to a eutrophic condition with the addition of phosphorus. Examination of Figure 6 reveals a rather peculiar graphic, one that is without the usual *x*- or *y*-axis. In effect, it is a normal graph which has been turned 90° to the left and is now upside down. The new vertical axis refers to depth, usually measured in meters, beginning with the surface at the top of the graph and continuing downward to the bottom of the lake. The new horizontal axis at the top of the graph represents a continuum for whatever parameter we choose to assess. For an oligotrophic lake, the two physical characters of greatest interest are temperature and dissolved oxygen, or DO. If these two parameters are measured in an idealized oligotrophic lake in the north temperate zone in the middle of summer, the profiles are striking. Temperature at the surface in July would be, let's say, approximately 22–25°C. If a thermometer is lowered from the surface of the lake to its bottom, the temperature remains at 22–25°C for the first 8 or 9 meters and then, abruptly, within the space of about 2 meters, it declines to 3.96°C (remember, we are dealing with a hypothetical lake under ideal conditions). One of the peculiarities of water is that it is most dense at 3.96°C, which explains why the cooler water is down deep and, coincidentally, why ice floats (ice is 0°C and less dense than at 3.96°C). The sharp decline in temperature coincides with a rapid change in its density, creating both a thermal and a density gradient. The density gradient is so powerful that warm water at the top of the lake and cold water below are completely separated and cannot mix. It is almost like there are two lakes within the basin comprising the lake. The thermal/density gradient is referred to as the thermocline. The area above the thermocline is the epilimnion and below is the hypolimnion. As can be seen in Figure 6, DO is also slightly different in the epilimnion and hypolimnion. Colder water and water under pressure will hold more oxygen, which accounts for the slightly elevated DO in the hypolimnion, and satisfies a couple of important gas laws.

Thermal stratification, which we see in our lake, will persist throughout the summer. However, as summer wanes, less and less infrared (thermal) radiation will strike the lake's surface. By some point in the late summer, heat in the lake will radiate into the atmosphere at a rate that is faster than it comes in, and temperature in the epilimnion will begin to fall slowly. As water temperature declines, the density gradient separating the epilimnion and hypolimnion will become progressively weaker. At some point, the gradient disappears completely and the lake undergoes mixis, or fall turnover. Temperature and DO at this time will be the same from top to bottom. If the lake is in the far north, it will acquire a layer of ice in the winter, which makes the temperature at the surface 0°C. If a hole is cut in the ice and temperature is measured from the surface down, a rise in temperature can be seen from 0°C to 3.96°C, and then it stays that way to the bottom of the lake. The lake is again thermally stratified (Fig. 7). As the season moves into early spring, increasing radiant energy from the sun will eventually cause the ice cover to break up, usually quite suddenly. An isothermal profile, similar to the one in the fall, then follows ice-off, and the lake undergoes a second mixis. As more and more radiation strikes the lake's surface, the water temperature begins to rise and the lake will

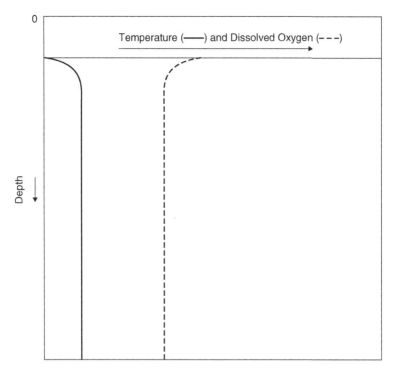

Figure 7 Thermal and dissolved oxygen profiles of an ideal north-temperate lake in mid-winter. (Courtesy of Joel Fellis.)

again stratify thermally, just as it did the previous summer. Lakes in the far northern hemisphere are dimictic because they turn over twice each year. In more southern latitudes, where a persistent layer of ice does not form in the winter, mixis occurs only in the fall and the lakes are monomictic.

In addition to possessing oxygen in the hypolimnion throughout the summer months, the oligotrophic lake has certain biological character-istics as well. For example, because of the relatively low water temper-atures, the lake will support what is called a cold-water fishery, com-monly including coregonids and salmonids in the north. These species are cold-water stenotherms, meaning they are able to tolerate only colder temperatures in a relatively narrow range. During thermal stratification, these species will, for the most part, be restricted to the deeper hypolim-netic regions of the lake. Primary production by phytoplankton and other plants in the lake is relatively low because nutrient loading, i.e., the phos-phorus level, is also low. The biota in an oligotrophic lake is generally species-rich, but the population sizes of those species that are present will be small.

Now then, let's perturb the lake by adding phosphorus. Humans have done this knowingly, and unknowingly, in many bodies of water throughout the world for hundreds of years. Phosphorus typically will come from nonpoint sources such as fertilizer run-off via farming operations, or from point sources such as faulty, or old, septic systems. Generally, the run-off is slow and gradual but, over a period of years, the acquisition of phosphorus produces the cumulative affect of cultural eutrophication. When phosphorus is added, planktonic algae will use it in photosynthesis and cellular respiration (ATP production), and for reproduction (DNA and RNA synthesis). Population sizes of many of these phytoplankton species will increase quickly. Mostly, these organisms are short-lived. On dying, they sink through the thermocline to the bottom of the lake where they are decomposed through the action of bacteria in the substratum. The decomposition process is aerobic and, as more and more phosphorus is added to the lake, the rate of oxygen depletion in the hypolimnetic water increases. Fortunately, at the time of fall turnover each year, water in the hypolimnion mixes with that in the epilimnion and the lake water again will become well oxygenated from top to bottom. However, when the lake stratifies the next summer, the hypolimnetic water is again "locked away" from an oxygen source until the fall mixis. Over a period of years, this process can, and many times does, induce complete anoxia in the hypolimnion (Fig. 8).

The impact of anoxia in hypolimnetic water will be devastating to the biota of an oligotrophic lake. In the first place, some of the phytoplanktonic species prefer eutrophic conditions and will out-compete those species that do well when the lake is oligotrophic. These latter species may become locally extinct. Benthic, or bottom-dwelling, organisms in the hypolimnion are generally cold-water stenotherms. As oxygen concentrations decline, they may be forced into shallower areas of the littoral zone where there is oxygen. However, migration into the shallow, and warmer, littoral water places them in competition with species that are better adapted to these conditions. Local extinction may certainly follow. The fate of organisms living up in the water column of the hypolimnion, also cold-water stenotherms, is similar to that of the benthic fauna when anoxia develops and they too may undergo local extinction.

During cultural eutrophication, the thermal profile in a lake will remain unchanged, but the DO profile is remarkably altered (Fig. 8).

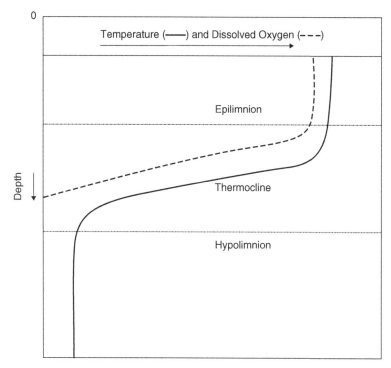

Figure 8 Thermal and dissolved oxygen profiles of an ideal north-temperate eutrophic lake in mid-summer. (Courtesy of Joel Fellis.)

However, if the addition of phosphorus is curtailed, cultural eutrophication can be reversed, and the DO profile will eventually return to that of the lake when it was oligotrophic. Obviously, this would be the wisest decision, to restrain the input of phosphorus, but destruction of the biota during cultural eutrophication is largely irreversible. In other words, local extinctions will, for the most part, be permanent. Yes, the cold-water fish species can be restocked, but not the phytoplankton, or the hypolimnetic benthic fauna, both of which may be totally eliminated during complete anoxia. Although the lake may again become pristine physically, it cannot be fully restored biologically.

The question before Wisniewski was whether eutrophication can affect the parasite fauna in a lake? Since the life cycles of many parasite species are dependent on complicated food-chain dynamics and predator–prey interactions, there should be an effect. But which species are most affected, or, indeed, are some more vulnerable to the shock of

Figure 9 Another perspective on Lake Druzno. This body of water became famous when Wincenty Wisniewski undertook his research on the relationship between eutrophication and host–parasite interactions. (Courtesy of Jerzy Rockicki.)

eutrophication than others? If so, which ones, and in what way? Does eutrophication impact host–parasite interactions? Many bodies of water in which the relationship between eutrophication and parasitism have been examined are shallow, including Lake Druzno (Fig. 9). Is Lake Druzno like it is in parasitological terms because it is shallow, or because it is eutrophic, or both? Will the depth of a lake, or the extent of the littoral zone, marginalize the impact of eutrophication? Will the parasite fauna in these bodies of water be affected by the trophic status of the lake, or by some physical characteristic that would alter the nature of the parasite fauna, irrespective of eutrophication? In a hypereutrophic lake, as we will see from Clive Kennedy's long-term study of Slapton Ley, winterkill can produce huge changes in the local biota, including the parasites. Winterkill occurs when a hypereutrophic lake is covered by ice. Light penetration may be prevented by a snow pack and, without light, phytoplanktonic photosynthesis will slow dramatically – even cease. Without replenishment of oxygen via photosynthesis, anoxia will develop and winterkill will occur.

As previously mentioned, Lake Druzno is situated at the northern end of the Elblag–Ostroda canal system. It is connected to the Vistula Lagoon via the Elblazka River. The lake is relatively large, approximately 10 km in length, and ranges from 0.5 to 2.0 km in width. Except for a channel dredged for the Elblag–Ostroda canal, the lake, however, is exceedingly shallow, about 1.0–1.5 m in depth on average. As a result, the littoral zone around the lake's edge is quite extensive, with large areas of both emergent and submergent vegetation. Wisniewski estimated that one-third to one-half of the lake's surface was covered by emergent vegetation from late spring to autumn during his study in the early 1950s. It is also an immensely important refuge for migrating waterfowl. Finally, the lake was eutrophic when Wisniewski conducted his investigation, and because of my own field experience with eutrophication and parasitism in several lakes in south-western lower Michigan of the USA, I became interested in Lake Druzno in Poland.

Wisniewski's efforts were extensive (Wisniewski, 1958). In 1950 and 1951, he and his group necropsied 348 birds, representing 50 of the 70 species occurring on the lake. A total of 373 amphibians from seven species was examined. Eight hundred and sixteen individual fishes from 21 species were necropsied. He also checked approximately 9000 molluscs, from 19 species. Other invertebrates included nearly 49000 microcrustaceans, 13000 insects, 17000 isopods, 20800 oligochaetes, and 10500 leeches. The dominant helminth parasites in Lake Druzno were tapeworms, with 67 species, and trematodes, with 84 species. From the avian hosts alone, 10664 individual trematodes, 41453 cestodes, 330 acanthocephalans, and 17 nematodes were recovered.

These staggering numbers, however, belie the most significant findings. Thus, from these data, Wisniewski formulated four very basic conclusions, three of which he asserted were generalizations that would have application for the parasite fauna in any body of water, and a fourth that he contended would apply solely to eutrophic lakes. First, he stated that the "final hosts of tapeworms, flukes and thorny headed worms are a sort of concentrating sieve for parasites in a water biocoenosis, while intermediate hosts serve mainly to help these parasites pass to their final hosts proper." Second, he indicated that parasites are not evenly distributed in various areas of a lake, "but occur in a greater congestion in some points." Hosts, he said, are similarly characterized by their heterogeneous distributions. Third, he noted that some parasites are typical of these systems and others are not. The dominant hosts within a lake will carry most of the typical parasites, e.g., in Lake Druzno the fauna was dominated by parasites that end up in birds, the ascendant definitive host group. Fourth, he stated that "in eutrophic bodies of water, particularly in shallow ones, the parasitofauna of birds prevails and is characteristic of them."

Wisniewski's second and third conclusions have been well confirmed since his 1958 publication. His second conclusion described the heterogeneous distribution of parasites and their hosts in aquatic ecosystems. There is now general agreement that most hosts are aggregated in their spatial distribution and, therefore, probably their parasites as well. Roy M. Anderson saw this in the distribution of oligochaetes infected with plerocercoids of *Caryophyllaeus laticeps* in the early 1970s, for example. Infected oligochaetes were found to be highly aggregated, or bunched, in their distributions. This overdispersed pattern of distribution in the intermediate hosts was then transferred to the piscine definitive hosts. The spatial distributions of larval trematode infections in mud-flat snails, *Ilyanassa obsoleta*, are also highly aggregated, as reported in a series of studies by Larry Curtis along the Delaware coast of North America. Julie Williams, in an investigation involving the hemiurid fluke, *Halipegus occidualis*, and its freshwater pulmonate host, *Helisoma anceps*, found that the snails were more likely to be infected in shallower areas of the littoral zone. The definitive hosts for the fluke are green frogs, which are ambush predators. They sit motionless in shallow water, waiting and watching for their prey. They also presumably defecate as they sit, shedding eggs of the parasite that are accidentally ingested by snail intermediate hosts as they graze in the shallow water. Since these snails do not move extensively, shallow water along the pond's edge is where the higher numbers of infected snails will be found.

In the same North Carolina pond in which Julie did her work, Derek Zelmer and Eric Wetzel conducted a series of mark–release–recapture studies on the green frog. They found that the frogs infected with *Halipegus occidualis* were mostly confined to four specific habitats in the pond, all of which were ideal for the parasite's intermediate and definitive hosts and, therefore, for transmission of the parasite. They related their findings to the idea of landscape epidemiology, a concept promulgated by several east European and Russian parasitologists, including Wincenty Wisniewski, E. N. Pavlovski, and V. A. Dogiel, in the middle of the last century. These are but a few of the many well-documented examples of the spatial heterogeneity that have been defined since Wisniewski's report in 1958.

Wisniewski's third conclusion is related to what is a "typical" parasite and what is not. He found that the overall parasite fauna in Lake Druzno was dominated by species that completed their life cycles in birds. Moreover, he observed that a number of avian parasite species were much more common than others. Wisniewski pointed out that "the typical species are at the same time more numerous and represented by more specimens, while those less typical are less numerous and represented by fewer specimens in their final, intermediate and first hosts." Hanski, in 1982, introduced, or shall we say, formalized, the concept alluded to by Wisniewski some 24 years earlier, although I am quite confident he was unaware of Wisniewski's earlier findings. Al Bush and John Holmes in 1986 applied the terminology developed by Hanski in their study of parasite component communities in lesser scaup. They described those species that are regionally common and locally abundant as core species, the so-called typical species of Wisniewski in 1958. Those that are regionally uncommon and locally rare are satellite species, or the "less typical species" of Wisniewski.

The recognition of heterogeneous distributions of hosts and parasites in Lake Druzno, and Wisniewski's codification of typical and less typical species were far-reaching ideas, way ahead of their time. For me, however, his first conclusion represented what was to become the future direction of parasite population ecology. Here, he referred to the final hosts of cestodes, trematodes, and acanthocephalans as "a sort of concentrating sieve" and gave emphasis to "the importance of various hosts in their development cycle" as "one of the most important features of the circulation of parasites in water biocoenosis." He went on to state, "This role of final and intermediate hosts has become clear only against the background of the whole biocoenosis and relations established by quantitative examinations." This last point, "by quantitative examinations," was crucial. It is quite unfortunate that he published this paper in a regional journal. He was actually among the first to recognize the importance of a quantitative approach to parasitism, almost 15 years before Harry Crofton published what are now considered as the seminal papers on quantitative parasite population ecology. During the August 2000 interview that I had with

Clive Kennedy, he remarked, almost wistfully, that Crofton's 1971 contribution was significant because "it placed numbers on life cycles." It seems to me that Wisniewski was saying the same thing 15 years earlier, though not quite with the same elegance or at the same level of sophistication Crofton was to provide subsequently.

My only real problem with Wisniewski's report of the work he did in Lake Druzno was in his last assertion, the one regarding the dominance of avian parasites in eutrophic systems. To reiterate, he stated, "in eutrophic bodies of water, particularly shallow ones, the parasitofauna of birds prevails and is characteristic of them." Since 1958 and the publication of his study, a substantial number of papers have been produced which deal with the relationship between the nature of a lake's trophic status and its parasite community. It is clear, to me at least, that most of the lakes in which these studies were done have an extensive littoral zone, irrespective of the lake's trophic status, thereby creating habitats which are attractive to a variety of aquatic birds. These shallow lakes likewise support a species-rich community of invertebrates, which serve as intermediate hosts for a wide range of trematode, cestode, and acanthocephalan parasites. This generalization may be tempered by a number of factors, e.g., the encroachment of human habitation, the availability of calcium in the watershed, and the geographic location of the lake, among others. I seriously question, however, that eutrophication is a necessary factor in creating a species-rich community of parasites, or birds, or invertebrates, or any other group of hosts. Indeed, if eutrophication leads to reduced species-richness in an oligotrophic lake, should we not expect the parasite fauna to be affected in the same way? In contrast, it is reasonable to expect a large and deep body of water, with a small littoral zone, to have a disproportionately less rich parasite fauna, especially among species that cycle through aquatic birds. It seems more plausible that the fauna in such a lake should be dominated by parasites that complete their life cycles in fishes, not birds.

My own experience with Wintergreen and Gull Lakes in south-western lower Michigan in the late 1960s supports this notion. Thus, for example, hypereutrophic Wintergreen Lake is very shallow, with an extensive littoral zone and a dominant, avian parasitofauna. Nearby Gull Lake is deep, and does not have much of a littoral zone. At the time we began working in Gull Lake, it was dominated by a piscine parasitofauna. Interestingly, even though we did not know it at the time, eutrophication had begun in Gull Lake, just prior to when we also started a long-term study on *Crepidostomum cooperi*, an allocreadid fluke that uses centrarchid fishes as the definitive host, sphaeriid clams as first intermediate hosts, and the burrowing mayfly *Hexagenia limbata* as the second intermediate host. Over a period of 20 years, we followed the population biology of metacercariae in the mayfly, during which time eutrophication was occurring, and then after it was reversed. The relationship between eutrophication and the population biology of *C. cooperi* in the mayfly was striking. The eutrophication process produced anoxia in deep parts of the lake, which, we hypothesized, drove the mayfly nymphs into shallow water where they more easily recruited the parasite's cercariae. A few years after nutrient loading was effectively halted, the population densities of the metacercariae rapidly plummeted in the mayfly, we suggested because they were able to return to their preferred habitat in the deeper parts of the lake. Otherwise, the parasitofauna in Gull Lake was unaffected by the eutrophication process, or its reversal, at least as far as we could tell.

In today's jargon, parasites that complete their life cycles in avian hosts that are associated with aquatic ecosystems are referred to as allogenic species (see also Chapter 2); these parasites do not complete their life cycles within hosts that are permanently confined to a lake, pond, or stream. Autogenic species include those whose hosts are confined to aquatic systems. Eutrophic systems, such as Lake Druzno and Wintergreen Lake, are typically dominated by allogenic species of parasites, but, I believe, not because the lakes are eutrophic. As noted previously, it seems much more reasonable that the dominance by allogenic species in these lakes is because they are commonly shallow and possess extensive areas of littoral zone. Deeper lakes, regardless of their trophic status, generally have proportionately smaller littoral zones and are usually dominated by autogenic species that employ fishes as definitive hosts. So, even though Wisniewski observed dominance by avian parasites in Lake Druzno, I would contend that he was probably incorrect in his conclusions regarding a relationship between the dominant avian parasitofauna and eutrophication.

As I have said previously in this chapter, in my judgment, Wincenty Wisniewski was far ahead of his time in terms of his research in ecological parasitology. In his pre-war years, he established himself as a solid young parasitologist, but World War II changed his interests and he began to think about parasitology in ecological terms. I found his eutrophication paper quite by accident when I read Dogiel's *General Parasitology* (1964), and saw it referenced there. I can say that Wisniewski's paper greatly influenced my early thinking about the dynamics of host–parasite interactions in aquatic ecosystems and, subsequently, my own development as an ecological parasitologist. I found it of great interest that he was among the first to employ the concept of spatial heterogeneity to parasites in aquatic systems and, as near as I can determine, the first to classify parasites as "typical" or "less typical" almost precisely the way in which we currently recognize core and satellite species. I was also impressed with his early emphasis on the importance of quantification which, subsequently, was to become the central theme of Harry Crofton's significant contribution in 1971.

This may seem like a real stretch, but because of this interesting man and his truly perceptive ideas, I was able intellectually to connect my Elbing in Kansas with his Elblag in Poland.

References

Ambrose, S. A. (1996). *Undaunted Courage*. New York, New York: Simon & Schuster.
Dogiel, V. A. (1964). *General Parasitology*. London, UK: Oliver and Boyd.
Wisniewski, W. L. (1958). Characterization of the parasitofauna of an eutrophic lake. *Acta Parasitologica*, **6**, 1–63.

2

Slapton Ley, and other matters British

Come forth into the light of things. Let nature be your teacher.

WILLIAM WORDSWORTH, *THE TABLES TURNED* (1798)

As I mentioned in the Prologue, I first met Clive Kennedy (Fig. 10) in the spring of 1972 when I was at Imperial College working with Professor J. Desmond Smyth on the in vitro culture of *Taenia crassiceps*. I had previously known of Clive because of his early parasite ecology work in the River Avon and at a small freshwater pond, called Slapton Ley, in the south of Devon in England. In corresponding with Clive during the course of writing this essay, I should note that he took umbrage with the manner in which I referred to his Ley. At one point he said, "You refer to the Ley as a small freshwater pond. In comparison with the Great Lakes, it is, but I think this may also be a matter of different usage across the Atlantic. We would call it a lake, and indeed it is normally described and referred to as the largest freshwater lake in the south-west" of England. On the one hand, I guess there is a difference in usage between the two sides of what some refer to as the "big pond." So, I will now defer to Clive's opinion and refer to the Ley as a lake (although, deep down, I still think of it as a pond). There are two points, on the other hand, about which we totally agree with respect to the Ley. First, it is a freshwater system and, second, it is a beautiful body of water!

While working in London at Imperial College in 1972, I took a train from Paddington Station down to Exeter where I spent most of the day talking with Clive about parasites and ecology. Later in the spring, I was to meet up with him in the English Midland city of Loughborough at the annual meeting of the British Society for Parasitology (BSP), where we continued our discussion about the ecological aspects of host–parasite

Figure 10   A photograph of Clive R. Kennedy at an acanthocephalan workshop
we attended in the spring of 1989 at a conference center in Exeter.

relationships. After I switched my research over to ecological parasitology on my return to Wake Forest later in 1972, I began corresponding
with him on a regular basis. We became close friends, and still are for that
matter.

Kennedy is an interesting man. I consider him to be one of the most
knowledgeable ecological parasitologists in the world, but he is far more
than a fine scientist. His interests are wide, ranging from the history and
politics of Britain and Ireland, to rugby. I recall my wife and I visiting
him in 1980 when I was invited over to the University of Exeter to serve
as the External Examiner (Viva Voce) for one of his graduate students,
John Aho. John had received his B.A. in Biology at Wake Forest, and his
Master's degree working with Whit Gibbons and me down at Savannah
River Ecology Laboratory (SREL). He had then secured a prestigious Full-
bright Fellowship to do his Ph.D. with Clive at the University of Exeter,

Figure 11   John Aho (in the foreground), with Clive (in the center), and me, on a site visit to the River Swincombe up on Dartmoor in the fall of 1980. This was the location of John's Ph.D. research and, as you can see, the wind was blowing hard (and it was cold!). (Courtesy of Ann Esch.)

and that is where I came back into John's professional life, as his External Examiner. Clive made arrangements for my wife and me to be housed in their magnificent faculty club, Reed Hall, on the lovely campus of his university. I not only remember the splendor of the physical surroundings, but the fantastic English breakfasts served by the staff each morning to my wife and me. We loved the pampering!

While in Exeter, Clive drove all of us up to Dartmoor (Fig. 11) where John was doing his research on *Cystidicloides tenuissima*, a nematode parasite that cycles through salmonids in the River Swincombe. During our drive to John's field site, Clive entertained us with ghost stories of the moor, also reminding us that Dartmoor was the back-drop for a Sherlock Holmes novel by Sir Arthur Conan Doyle, *The Hound of the Baskervilles*. What an absolutely stark environment up on the moor! There are no trees up there, it rains a lot, the wind blows, it rains a lot, the vegetation is mostly gorse and coarse grass, it rains some more, and the wind blows, a lot. Later that same day, Clive took us to what was supposedly a sixteenth-century pub in a tiny village near the moor. (A question about English pubs: Why were so many of them seemingly built in the sixteenth or seventeenth century? It isn't that I don't believe their claims, I just think it interesting that almost every village in England has one that was built about that time.) At any rate, it was at this pub that my wife and I were introduced to the well-known Ploughman's lunch, a refreshing combination of cheese, homemade bread, lettuce, tomato, maybe a pickle, and perhaps some slaw or potato salad. We relaxed and talked about Dartmoor and John's work as we ate our lunches and enjoyed a pint of bitter sitting in front of an open fireplace and a warm blaze. It was a wonderful interlude in a fantastic old pub, regardless of whether it was built in the sixteenth century or not.

On an evening later in the week, we were guests of Clive and his wife at their thatched-roof home in the tiny village of Bow, which sits at the edge of Dartmoor. What a grand time we had, and an absolutely sumptuous meal. It was our first close-up experience with a home-cooked English meal. We had always heard about how bland English food was supposed to be, but definitely not so if it is prepared the way his wife did it, e.g., delicious roast beef and Yorkshire pudding. We had seen many thatched-roofed English cottages, but had never been inside one. Low ceilings and small rooms, to wit, very cozy. In addition to the rest of his wonderful home, Clive showed us the secret hiding place, a priest's hole, in back of the fireplace. The thatched-roof home was built in 1560 during the reign of Elizabeth I. Throughout much of this period, Protestants persecuted Catholic priests, and the well-hidden priest's holes provided refuge. These secret nooks and crannies were used again to hide Catholic priests who were on the run during the English Civil War, which lasted from 1642 to 1646. Many parts of western England had remained loyal to Charles I in his dispute with Cromwell and his Roundheads, the latter so named because of the peculiar way of cutting their hair, and a statement of contempt for the royalists who wore theirs at shoulder length. Unfortunately for Charles, he was to lose, first the war, and then all of his hair, along with his head (outside the Banqueting Hall) in London in 1649. Regrettably, Clive and his wife decided to sell their magnificent old home not long after our visit. Clive confided that they were much too restricted by all the regulations placed on them, as the structure was considered part of British national heritage and they were unable to do any sort of remodeling as a result.

After that visit, Ann and I returned to Exeter many times over the years. We enjoy the ambience of the city, with its wonderful quay and shops along the River Exe. There is also the absolutely magnificent St. Peter's Cathedral, the construction of which began in the eleventh century. We can remember attending a beautiful Evensong service in the Anglican Cathedral the first time my wife and I visited Exeter, and then a Christmas musicale on another visit. Right across the square from the Cathedral is a truly fantastic restaurant called The Ship. A tavern in its early days, it was said to be a watering hole for Sir Francis Drake. Exeter was an open port for shallow-draft ships before the River Exe silted up. It seems, according to Clive, that the Countess of Devon built a weir (which still exists as the Countess Weir) to stop ships from sailing up the Exe because of a dispute she had with the city of Exeter and its right to tax her properties. In retaliation, the city authorities built a canal to bypass the weir. A problem that then emerged was silting of the river because of the weir. Even with the increased size of the ships and the severe silting, however, Exeter remained a thriving port to the end of the nineteenth century. Clive remembers when he first arrived at the university that timber ships still sailed up the canal and berthed at the city's quay. Whatever, each time we visit Exeter, we convince Clive that we need to go to Drake's place and enjoy one of their delicious rump steaks, along with English peas, chips, and, of course, a pint of their fine lager.

On a trip to Exeter in 1987, we were joined by Al Bush, an old and dear friend who teaches at Brandon University in Brandon, Manitoba. This is where I met Cam Goater the first time as well (Cam is the brother of Tim Goater, one of my Ph.D. students and an undergraduate student of Al Bush). Cam also had done his undergraduate work with Al, so it was a fun time for all of us. On this excursion, Clive managed to secure lodging for us again in Reed Hall, where we were spoiled again by the superb service and the elegant surroundings. One reason for going to Exeter on that occasion was not only to visit Clive, but to do some last-minute preparations for papers that all of us were to deliver at the BSP Silver Anniversary meeting in Edinburgh, Scotland. Before heading up for the meetings in Scotland, Clive took us out in the field to see his study site at Slapton Ley for the very first time, down on the coast of Devon. I was really impressed with the place. A painting of Slapton Ley by our daughter, Lisa

Figure 12   Slapton Ley on the coast of Devon, England, the site of Clive Kennedy's long-term research. This photograph was taken high above the Ley, and adjacent to the cliffs where the Ranger battalions trained for their climb of Pointe du Hoc at Normandy. The Slapton Sands are to the left of the Ley, and then the English Channel.

Esch McCall, graces the cover of this book. It was done from a photograph which I took on that particular trip.

The lake is on the small side, freshwater, and is situated about 100 meters from the sandy beaches of the English Channel. The lake is of interest for a couple of reasons. First, Clive had used it for many years in the conduct of his research, about which I will talk more later in this essay. But, it is also significant because of its location near an important World War II historical site. When one approaches Slapton Ley (Fig. 12) from Exeter, it can be seen initially from some fairly high ground to the east. Adjacent to this high ground there is a stretch of cliffs and sand beaches along the Channel. These cliffs were used by the Second and Third Ranger Battalions to rehearse for their climb of Pointe du Hoc between Omaha and Utah beaches at Normandy on D-Day, 6 June 1944. The Rangers had been assigned the task of knocking out several large coastal batteries which, supposedly, had just been installed by the Germans. Unfortunately though, the intelligence that reported the presence of these guns was incorrect as they were still some 2 miles inland and not yet operational. Ranger losses were heavy during their relentless and successful, but needless assault. Not only were the cliffs used for rehearsing the landing of the Rangers, most of the beaches in the vicinity of what is called the Slapton Sands were used to practice for the D-Day landings. Allied reconnaissance teams had been secretly sent by submarine to the Normandy beaches where they collected sand to see if it would support tanks and other heavy equipment necessary for the success of the planned landings in early June of 1944. They found the beaches at Slapton Sands were virtually identical to those at Normandy. Tragically, a German E-boat, similar to an American PT boat, managed to sneak in among allied transports and sink one of them, with the loss of nearly 800 men and all their equipment. A few days before Clive took us on our excursion to Slapton Ley, we saw on BBC television the dedication of an American Sherman tank, recovered from the Channel, as a monument to the brave men lost in that action.

Our travel to Exeter and Slapton Ley was part of a pleasurable sojourn to the UK in that spring of 1987. Ann and I took an overnight train from Exeter via Bristol to Edinburgh for the BSP meetings. That too was an interesting part of our overall trip. We boarded the train in Bristol about 10:30 at night, found our compartment, and went to bed. However, we were awakened by a knock on the door in the middle of the night and instructed by a uniformed officer that we were to move immediately to the car behind us on the train. Moreover, we were told to leave our suitcases, that someone would be along shortly to transfer them to our new compartment. We asked why we were being forced to move so abruptly. He responded only by saying, "there was a problem." We speculated many times afterwards about why they wanted us to move and, of course, our reasons became more and more exaggerated as the years have passed. I think the last one had us in the middle of a drug-smuggling ring and Scotland Yard, or something like that.

During our stay in Edinburgh, Clive took Al Bush, Rachel Bates (one of his graduate students), Ann, and me on a guided tour of Edinburgh Castle and then on a walk down the Royal Mile to Holyrood House. The latter is the charming Royal Palace, the Queen's official residence in Scotland. The castle at the south end of the Royal Mile is both imposing and impressive. It was while we toured the castle that Clive's historical proclivities really came to the fore. As we walked, we were regaled by tales of historical England and Scotland, of Edward I (Longshanks) and William Wallace (long before Mel Gibson and *Braveheart*), Edward II and Robert the Bruce, and, of course, Mary Stuart, Queen of Scots. At one point during our tour, I can vividly recall coming to a large stone building and stopping in front. Clive asked if we wanted to go inside, and we followed him in. We entered a massive hall, with many flags and other heraldry adorning the walls. He immediately turned to his left, as though he knew exactly where he wanted to go. He went straight to a table on which were a number of very large books. He found the one he wanted, opened it to a particular page, and pointed to a name: it was Kennedy. He then explained that it was his father's and that he was aboard a merchant ship that had been torpedoed by a German U-boat in the north Atlantic just prior to Clive's birth in Liverpool in 1941. The books on the tables contained the names of all the Scots who had been killed in World War II. We were in the building that served as Scotland's national war memorial. Ever since I began reading about Poland, Elblag, and Lake Druzno, I have wondered if the submarine that sunk his father's merchant ship had been built in Elblag.

Our BSP meeting in Edinburgh was fabulous. We were able to see many of our old friends and make several new ones. Without question, though, the highlight of the meeting was their evening banquet and being piped into the large dining room by a giant of a man, dressed in a kilt, with a full red beard, and looking every bit the Scotsman we were certain he was. The meeting was over on a Friday, so Ann and I took a train up to Inverness where we stayed for a couple of nights at the venerable old Station Hotel. If you ever get to Inverness, a visit to this hotel is well worth the time. Although the room in which we stayed was quite small, the overall atmosphere of the hotel reflects its setting in the Scottish Highlands. It is absolutely charming. The staircase leading up to the second floor from the foyer is one of the most beautiful we have ever seen, and the food was exquisite in the elegant dining room. Our excuse for going all the way up to Inverness was, of course, a visit to Loch Ness and, as every tourist hopes, perhaps to catch a glimpse of Nessie. So, the next morning we walked around the corner and picked up our rental car. Even though we had visited the UK a couple of times, I had never before driven on the other side of the road. Without going into all of the excruciating details, the first thing I did was to get lost in a residential neighborhood far away from the highway leading to Loch Ness. How I got in there, I still do not know. Finally, after driving around for about 30 minutes, still completely lost, and with Ann nearly in tears, we decided it would be best to pull into a parking lot and check the map to get our bearings. As I turned in, the first thing I saw was a very large truck coming straight for us! I again was on the wrong, or right, side of the entrance. Fortunately, we were in a parking lot and neither of us was going very fast, so we avoided a collision, but the driver of the truck sure gave me a dirty look as he passed by, with me still on the right-hand side of the road.

Ann was ready to go back to the hotel and skip Loch Ness, but I persuaded her we could still get there. After checking our map and with Ann now navigating, we finally found the highway and made our way to Fort Augustus at the south end of Loch Ness. We had to drive through some rather hilly terrain, with some still substantial patches of snow on the ground. After all, we were in the heart of the Grampian Mountains and it was early spring. We made it without incident. Despite our previous bad experiences, the visit to Loch Ness was very much worth all our earlier agony. I even managed to obtain a couple of interesting photographs (Figs. 13 and 14) of Urquhart Castle, which had been destroyed during the Jacobite uprising of 1745. That was about the time when Bonnie Prince Charlie and some of his Highlanders scared the "what for" out of the English by invading all the way down to Derby. At this point he retreated, then was eventually cornered and brutally crushed at nearby Culloden in

Figure 13 Urquhart Castle on Loch Ness, Scotland, during a March snow shower. My wife and I visited this site immediately following the Silver Anniversary Meeting of the British Society for Parasitology in Edinburgh, Scotland, in the spring of 1987.

Figure 14 Urquhart Castle on Loch Ness after the snow shower had passed over. The white stuff across the Loch is snow falling on the other side (5 minutes earlier (Fig. 13), it had been falling on the castle and us). The weather in Scotland at this time of the year is capricious, to say the least.

1746, courtesy of the Duke of Cumberland, son of George II. I am quite pleased with these photos, which is my only reason for including them here. By the way, Figure 13 was shot from inside our rental car through a split-rail fence where we had parked during a heavy snow shower. Five minutes later, after changing the film in the camera, the snow had stopped and the sun was out, so I got out of the car and snapped the Figure 14. The white patch above Loch Ness on the far shore is the snow shower that had passed over us 5 minutes earlier. Quite a place, Loch Ness, very impressive. But, alas, we did not see Nessie, the monster, although we certainly looked hard for it. We finally made it back to Inverness where we returned the car, promising ourselves never to drive in the UK, or Ireland, again. We broke that promise just one time when we rented a car in Galway City in Ireland and drove down, without incident, to the southern end of the Dingle Peninsula just before heading for the August 1990 International Congress of Parasitology (ICOPA) meetings in Paris. We had no trouble in Ireland when I drove the second time, so I guess the Scotland experience taught me enough about driving on the wrong (right) side of the road.

As I mentioned earlier, our trips to the UK and Ireland have been numerous. Because of these excursions, Ann and I became unabashed anglophiles. We generally manage to include visits with several old parasitology friends such as David Rollinson, Rod Bray, Vaughan Southgate, and David Gibson, at the British Museum of Natural History in London. Each time we stop at the museum, we usually have tea with the crew and then enjoy a wonderful lunch at one of the numerous sixteenth-century pubs near the museum. The setting for tea at the museum is generally way up at the top of one of the twin towers which are on either side of the front entrance to the building. One of our museum friends told us that the British equivalent of the American Secret Service always places a sharp shooter in his office high in the same tower whenever Prince Charles attends a Board of Governors meeting at the museum. We also frequently saw Phil Whitfield and would have a pub lunch in the Kensington area close by Kings College where he used to teach. On another trip we met up with Richard Tinsley after he moved to the University of Bristol where he was, until recently, Head of the Zoology Department, and Chris Arme hosted us at Keele University. On a couple of our excursions up to Scotland, we would see David Crompton at the University of Glasgow. I am fascinated by the museum at the University of Glasgow because of the large collection of paintings, etchings, and memorabilia bequeathed to them by James Abbott MacNeil Whistler. My middle name is Wisler, the spelling of which was changed at some point from Whistler, although we have never been certain of when or why. Whistler's biography indicates that the MacNeil side of his family immigrated to North Carolina from the Isle of Skye in 1746, not long after Culloden. The MacNeils had supported the Bonnie Prince and were persona non grata after the Highlander massacre at Culloden. During one of these trips, we also discovered the small village of Dalry, from where my wife's grandfather and his parents had emigrated to the USA in 1865, settling first in Iowa, and then (where else?) near Elbing, in Kansas.

We have also traveled over to Ireland several times. On a trip to Trinity College in Dublin, Celia Holland entertained us with lunch at their lovely faculty club. I remember sitting with Celia after lunch and discussing a research protocol, which was ultimately to lead to a doctoral dissertation for one of my graduate students, Derek Zelmer, who now teaches biology at Emporia State College in Kansas, not really that far from Elbing. Our time with Kieran McCarthy at University College Galway included a trip out to beautiful Loch Corrib and then a wonderful sail on the boat of one of his graduate students. A few years ago, we visited Mick O'Connell on the stark, but absolutely spectacular, Aran Islands in the mouth of Galway

Bay, and, yes, we saw "the sun set on beautiful Galway Bay." The journey out to the Aran Islands included an hour-long ride on a rolling ferry boat. Fortunately, the sea was relatively calm both in the morning and afternoon, and we did not lose either breakfast or lunch. I first began corresponding with Mick when he sent me a reprint request and I became curious about this person doing parasitology out on these remote islands. I learned that he was completing his Ph.D. at University College Galway and that he was doing parasitology as an aside to his primary research on the ecology of sand eels. I also discovered that Mick is an extraordinary naturalist as we toured the largest of these islands. It was in the fall, but the countryside was covered with many varieties and sorts of plants, most of which seemed to be in bloom. Mick knew them all by name.

In the summer of 2000, my wife Ann and I managed to wangle an appointment as resident directors of Wake Forest University's Worrell House at 36 Steele's Road in the Hampstead area of north-central London. What a great place to be! Our university uses the house as a residence for 15–16 students each semester who go over to be taught by a Wake Forest faculty member. In the summer, for 5 weeks, the house is sort of an inn for faculty, staff, and alumni who might want to stop and stay inexpensively for a few days while in London. It was during this trip that I arranged to take a train down to Exeter again and visit with my old friend, Clive Kennedy. He was ready for me and we spent four or five hours talking about a wide range of topics, but mostly about his life and his work at Slapton Ley.

I already knew that Clive was born in Liverpool in 1941, on 17 June. His father, a native of Paisley, had sailed from Glasgow on 13 June on a merchant ship that was torpedoed and sunk in the North Atlantic within a few days of Clive's birth. Clive was raised in Liverpool and, as with many biological scientists, he had originally thought about a career in medicine, but was drawn into the academic arena as a zoologist. He told me that the Head of his school had wanted him to become a physician and go to Oxford to study biochemistry, on his way to becoming a physician. When he announced his decision to attend the University of Liverpool and work toward a degree in zoology, his Head of school was furious, roundly denouncing his choice of university as a "technical school"! It is interesting that, despite the tremendous efforts of Thomas Huxley, Joseph Hooker, and others in the nineteenth century to help develop the reputations of universities like the one at Liverpool, Oxford and Cambridge (Oxbridge) continue to dominate higher education in the UK, even into the twenty-first century. While at Liverpool, Clive was trained in classical zoology, with an emphasis on aquatic, and invertebrate, biology. His first experience with parasites came when he took a course in parasitology from Jimmy Chubb during his fourth, or honors, year. Interestingly, this was also the first time Jimmy had ever taught the course, and the two were to become lifelong friends and colleagues.

Although Clive was fascinated by parasites, the "hook" was not quite set. When he decided to go for his Ph.D., again at Liverpool, he applied for a fellowship from the Nature Conservancy to support his graduate work. He told me the story of how intimidated he felt by the presence of the revered Charles Elton who served on his interview committee. During the questioning, he was made almost speechless when suddenly Elton asked, "do the seagulls still roost at Sefton Park in Liverpool"? He later concluded that Elton must have lived in the same area of the city and bicycled the same route that Clive used on his way to school each day. He apparently impressed Elton because he was successful in his application for graduate support from the Nature Conservancy. His doctoral dissertation dealt with the biology of several species of *Limnodrilus*, tubificid worms. During this research, he discovered that his tubificids were infected with both *Archigetes* spp. and larval caryophyllaeids and, as a result, about 40% of his dissertation was devoted to these most interesting cestodes.

As with all of us when we complete our journey in the "womb" of graduate school, usually with the support of an understanding mentor, we are required to face reality, i.e., the real world. I was interested to find that Clive's first employment was at University College Dublin, where he completed the writing of his dissertation and submitted his Ph.D. Then, in the following year, he moved to the University of Birmingham and another temporary job. The next year, however, three permanent positions came open, one at Aberdeen, another at Durham, and the third at Exeter. The position at Exeter called for an aquatic ecologist/parasitologist, an ideal position for his interests and talents. After his interview – the only one he had – he was offered and accepted the job at Exeter where he spent the last 35 years of his career. Seven of his last eight years were as Dean of Science and Head of the School of Biological Sciences. Conditions in British higher education were rather difficult during the 7 years he served in this administrative post and, though I suspect he would not say it, my guess is that this period in his academic life was the least rewarding personally and the most demanding professionally. Government "bean counters" and some academic administrators have a diabolical way of creating all sorts of nasty hoops through which departmental chairs are required to jump, no matter whether in the USA or the UK. In some ways, though, I have the impression that the education bureaucrats in London are even worse than some academic deans I have seen in action here in the States. Having been a regular faculty member (and still am), a department head (we call them chairmen), and a dean (but a good dean, not a wicked one), I have enough experience to know justifiable "burn-out" when I see it. Unfortunately for the discipline of parasitology, my good friend Clive Kennedy chose early retirement on completion of his administrative duties. As of 1 July 2001, Clive became Professor Emeritus, and is no longer an active member of the academy.

Whereas Clive spent a great deal of his early research time in aquatic ecology and fisheries biology, he gradually moved more and more into parasitology. Slapton Ley was to become the first real focus of his research dealing with host–parasite relationships, one which was to last throughout his career. His love affair with the Ley, as he calls the lake, actually began quite by accident. When he took the position at Exeter, his new department taught a field course in which all members participated. Unhappy with the field site they were using when Clive arrived, they eventually decided to employ a facility operated by the Field Studies Council at Slapton Ley (Fig. 12), and this experience was the real beginning of Clive's long association with the lake.

According to Clive, "Slapton Ley is a freshwater lake, created by the natural damming of a small stream, the River Gara. It is not open to the sea and water flows out through a single bar." Local geologists speculate that there were originally seven lagoons, which had been present for at least a thousand years along that stretch of the Devon coast. Four are now gone. Originally, there were two lakes that comprised Slapton Ley, but in the nineteenth century a road was built, diverting water from one of the lakes into the second. During winter storms, the outlet to the Channel may become blocked with shingle and sand, and the outflow can be stopped, causing the Ley to flood, much to the consternation of the local farmers. I asked Clive if the lake ever became brackish since it is so close to the Channel. He explained that, "on occasion, a good storm could move seawater into the Ley but, within 2–3 days, it would be gone," so the volume-replacement time is fairly rapid. The total surface area of the Ley is about 80 ha, the maximum depth is about 2 m, and the catchment basin is approximately 42 km^2.

His initial parasitology research in the Ley began rather by accident. It seems that a warden told him the local fishermen were complaining about the unusually small pike, roach, and rudd being caught in the Ley, and why didn't he have a look at the problem. So, he did, with the help of a National Engineering Research Council (NERC) grant. His original objective was to develop a management scheme to restore the fisheries. When he began his work, he hypothesized "that the predatory pike were small because the roach and rudd were of uniformly small size and that, energetically, there was minimum 'pay-off' for the pike in consuming the smaller fish." However, at some point during those first few years, the pseudophyllidean cestode, *Ligula intestinalis*, colonized the Ley, a fortuitous event, as he was soon to discover. The larval form of this parasite can produce devastating consequences to a local fish population when it appears. Previous studies by Chris Arme and others had shown that larval *L. intestinalis* apparently produces a chemical that interferes with the hormonal interactions involving the gonadal–pituitary axis, resulting in both sterilization and stunting of the fish even as the plerocercoid continues to grow in size. At the host population level, sterilization results in the elimination of the fish from the gene pool, in effect producing reproductive death. As the biomass of the plerocercoid continues to increase, the fish's swimming behavior is altered, significantly increasing the chances of the piscine intermediate host being consumed by the avian definitive host. Because of *L. intestinalis*, he speculated that the roach population size began to decline, in the same way that Crofton's 1971 model predicted it should. Clive told me his final report to the NERC indicated the idea of "managing the fish population in Slapton Ley was going to be impossible, due to the colonization of *L. intestinalis*."

After this initial experience in the Ley, I asked Clive if, at the time, he had any intention of working there for the next 25 years or so? His reply was an emphatic "no!" However, serendipity is a common experience of many who have worked in a single system for a long period of time, and a serendipitous event was to occur in Slapton Ley in 1976, something that was to keep his attention on the lake, even to his retirement. The lake underwent eutrophication; indeed, Clive says that it became hypereutrophic. This unexpected occurrence was precipitated by several events. The first was the imposition by the European Union of new regulations that required at least some of the local farmers to shift their focus from dairy herds to grain production to remain profitable. This brought about the increased use of fertilizers in their farming operations, and, with that, the run-off of phosphates and nitrates into the Slapton Ley catchment, then into the Ley itself. With the shift to grain crops, farmers began plowing their land. This led to run-off of topsoil, which enhanced silting in the Ley, causing it to become shallower. This exacerbated the eutrophication process. In 1976, there was a severe drought and this made eutrophication even worse.

The eutrophication that occurred in the Ley was to have a significant impact on the biology of the system. However, another long-term process also took place, one that contributed greatly to the overall changes that were to occur in the lake. In the early part of the twentieth century, grebes had been greatly decimated in the UK by commercial hunters who were interested in their plumage for the hat trade. Then, with the onset of World War II, pressure on these birds disappeared completely, as the British began shooting at Messerschmitts and Fokkers rather than grebes! It also seems there was considerable need in the war effort for the construction of roads, airfields, and the like. Such facilities required concrete, and concrete required sand and gravel. Consequently, large numbers of gravel pits suddenly appeared throughout southern England. An important consequence of the pits was the creation of superb habitat for the grebes, and their numbers rapidly increased as a result. In 1972, they were spotted at the Ley for the first time that anyone could remember and, in the next year, they began nesting. Within a few years, there were 18 pairs of these birds breeding successfully in the lake. The Ley was attractive to the grebes in part because it was an inviting habitat, but also because there was a large population of perch and stunted roach, ideal for feeding nestling birds. However, with the grebes came *Ligula*, as well as several species of eye flukes, including *Tylodelphys podicipina*. All of these parasites were to play an important part in Clive's long-term studies in the lake. And, with the introduction of these parasites into the lake, Clive's interests were also captured by the idea of parasite colonization. It was along about this time as well, that he, Al Bush, John Aho, and I began collaborating on a colonization paper in which we were to introduce the allogenic and autogenic species concepts. Our aim was to separate and characterize, in part at least, parasites capable of colonizing easily and those that were less able (likely) to colonize. Indeed, we used data largely generated in the UK and contributed by Clive, who also had the very major role in writing the paper that was ultimately published in *Parasitology* in 1988 (Esch *et al.*, 1988).

Anecdotally, I should note here that the original idea regarding allogenic and autogenic parasites came while Al Bush and I were sitting in the

San Francisco airport bar. We were waiting for a flight to take us home after we had both given invited papers at an American Association for the Advancement of Science (AAAS) symposium organized by Joe Schall back in 1975. Neither Al nor I are too fond of flying and, as a way of getting our minds in order before going up, he and I would frequently find ourselves sipping a beer, or a Bloody Mary, before climbing (sometimes almost crawling) aboard the aircraft. It was during the session in San Francisco that we began a conversation regarding the structure of parasite communities. We eventually involved Clive Kennedy and John Aho, and the 1988 paper in *Parasitology* evolved after 13 years of discussion in various airports, campuses, and pubs, in the UK, USA, and Canada.

After colonization of the eye flukes into the Ley, Clive became interested in the survival of *T. podicipina* metacercariae in perch over the winter months, and, accordingly, in January of 1985 he and his students set a number of gill nets in the lake. Unfortunately, that winter turned out to be one of the harshest in several years and the Ley acquired an ice layer that remained in place for nearly 3 weeks – a rarity for that part of England. The presence of ice on the lake prevented them from retrieving their gill nets in a timely manner. After the ice disappeared, they resumed efforts to collect fish, but were totally unsuccessful, even into late spring, at which time the anglers began complaining about the absence of fish in the Ley. After considering all potential possibilities, Clive concluded that the disappearance of fish from the lake was due to winterkill, which had been brought on by the eutrophication that had begun in the mid 1970s. Winterkills in eutrophic and hypereutrophic lakes in the north temperate areas of the world are not uncommon, and may be quite devastating to the local fish fauna. Although the local fishermen were clearly upset by the new state of affairs in Slapton Ley, Clive seized on the winterkill as an opportunity. With his growing interests in colonization as an important phenomenon in many host–parasite systems, the decimation of the fish and parasite populations in the Ley presented him with a chance to observe recolonization of the entire host–parasite enterprise in the Ley.

As in many cases following a period of disaster like the winterkill in Slapton Ley, or selenium poisoning in Belews Lake (see Prologue), there is a small community of fishes residing in a part of the lake or pond that is marginally affected by the environmental insult. These fishes, or those from feeder streams, represent the founding entities for the new fish community that will eventually become re-established. The process in Slapton Ley was surprisingly slow but, by 1989, the fish had begun to return in large numbers. And, as Clive said, "history repeated itself." The rudd came back first, followed by roach and, in turn, by a decline in the rudd numbers. The roach began to stunt. All but one of the parasites disappeared during the winterkill since their numbers dropped to levels below those necessary for successful transmission. The return of *Ligula* coincided with the reappearance of breeding grebes and, almost simultaneously, the eye flukes also returned. The system had become recast in a manner similar to the state in which Clive had found it some 25 years before. The presiding events leading the re-establishment of the new community were the drought in 1976 that led to hypereutrophication of the Ley, followed by the massive winterkill in 1984–85.

Another of the more interesting features of Kennedy's long-term association with Slapton Ley has to do with three species of eye flukes, i.e., *Diplostomum gasterostei*, *Tylodelphys clavata*, and *T. podicipina*, all of which occur in the humour of the eyes of perch, *Perca fluviatilis*. Elizabeth Canning and co-workers published the earliest parasite study of fishes in the lake in 1973 and, in that paper, reported the presence of *Diplostomum spathaceum*, but none of the other three species. This species, however, occurs in the lens of the eye and not the humour. Kennedy noted the presence of *D. gasterostei* and *T. clavata* for the first time in 1975. He suggested that colonization of the latter species had coincided with the appearance of great crested grebes at the Ley. Initially, the abundance of *D. gasterostei* remained stable, whereas that of *T. clavata* increased over time. Then, in 1976, the third humour-dwelling fluke, *T. podicipina*, made its appearance. As the abundance of *T. podicipina* increased, that of *D. gasterostei* began to decline. Clive believed that the latter species was declining in number because of competition with the two species of *Tylodelphys*.

The problem with this hypothesis was that it was not testable, for a number of reasons. However, the advantage of being persistent in monitoring the Ley over the long term was to become apparent to Clive as time passed. As was described previously, the European Union required the local farmers to shift their focus away from dairy production to grain. Accordingly, the use of fertilizers was increased and the Ley underwent eutrophication due to phosphate accumulation. Then, in the winter of 1984–85, there was a severe cold snap and the lake was covered by ice. Winterkill decimated the perch, rudd, and roach populations, along with their humour-dwelling parasites, except for *T. clavata*, which managed to hang on. Immediately following the fish crash, grebes significantly declined in numbers, as might be expected given the lack of forage fish. Clive seized upon this environmental "disaster" to conduct what he called a "natural experiment," i.e., to follow the parasite recolonization pattern and determine if his predictions regarding eye fluke competition could be supported.

As noted previously, recolonization by the fish species was slow, with very few being caught until 1989. By 1991, however, perch and grebe populations had returned to their pre-crash levels. *T. clavata* had continued to cycle in the lake, remaining at pre-1985 levels. By 1991, *D. gasterostei* had recolonized the Ley, but it did not return to its pre-crash abundance even though snails, fishes, and birds were there in sufficient numbers to support the parasite's life cycle. *T. podicipina* remained absent. "Why then," Clive asked rhetorically in his 2001 paper in *Parasitology*, "in the absence of *T. podicipina* did *D. gasterostei* not increase in density to at least 1976 and 1977 levels as might have been predicted by the competition hypothesis" (Kennedy, 2001)? His explanation rested with the notion that *T. clavata* had remained at pre-crash levels throughout the entire period and that it was "the only species whose population densities have been shown statistically and significantly to be inversely correlated with those of *D. gasterostei*." He went on to note that, whenever the two species were found together, the numbers of *D. gasterostei* were inevitably low. *T. podicipina* finally reappeared 1994, but did not regain pre-crash levels until 1999. Clive reasoned that the ability of *T. clavata* to emerge as the dominant species of the trio was based on its greater reproductive capacity and colonization ability. Persistence of all three species in perch eyes in the Ley was also possible because all three of them are overdispersed, or contagiously distributed, in the perch population. Because of overdispersion, a few hosts have many parasites, but most have a few or none. Those with small numbers represent situations in which the intensity of competition would be less, thereby allowing for successful recruitment and establishment of more than one species at a time. In his 2001 *Parasitology* paper, Clive concluded, "The extended data set and the recolonization of perch by the three species of eyefluke in Slapton Ley constituted a natural experiment that confirmed that negative interaction is the most parsimonious hypothesis that can be erected to explain the decline of *D. gasterostei* in the lake."

The eye fluke story is but one of the two that can be told within the context of Slapton Ley and Clive's long-term work in this wonderful place. The other concerns the pseudophyllidean tapeworm, *Ligula intestinalis,* as described in a 2002 *Parasitology* paper (Kennedy *et al.*, 2002). I have already alluded to certain aspects of this study and to the potential impact the lar-val stages of this pseudophyllidean may have on its second intermediate host, which is roach, *Rutilus rutilus*. As was explained earlier, he was ini-tially drawn to the lake by a local warden who told him about stunted populations of roach, rudd, and perch that were developing and asked if he would have a look at the situation. It seems that, with the onset of eutrophication in the late 1960s, the roach population increased dramati-cally, to the point that by the early 1970s, roach in the Ley had become stunted and interspecific competition between roach and rudd led to a decline in the hitherto previously dominant rudd population. It was at this point that the great crested grebe began colonizing the lake and with this piscivorous bird also came *L. intestinalis*.

Introduction of *L. intestinalis* into the lake resulted in even further dra-matic shifts in the dynamics of both the roach and rudd populations. As was noted, *L. intestinalis* not only stunts the growth of its fish host, it in-duces sterility, which may as well be host death as far as reproduction is concerned. As a result, roach numbers began to decline and rudd then proceeded to recover. Concomitant with the decline in roach densities, *L. intestinalis* also declined in abundance within the Ley. In 1984–85, the severe winterkill occurred and the densities of all fish within the lake de-clined precipitously. At this point *L. intestinalis* disappeared completely from the lake, along with its definitive host, the great crested grebe. As with the eye fluke population dynamics, Clive realized he had an oppor-tunity to conduct another "natural experiment" by following the recolo-nization of *L. intestinalis*, along with roach and rudd.

After the winterkill of 1984–85, rudd was the first to recover and, as in the late 1960s, it became the dominant fish species. However, as soon as the roach reappeared, rudd began to decline in numbers. Coinciding with the increasing roach populations, the great crested grebe returned. With the recolonization by grebes also came *L. intestinalis*, which then in-creased in density as it had during the first cycle in the 1970s. Moreover, as *L. intestinalis* increased, the roach population began to decline, dupli-cating the pattern in the first cycle. As might be guessed, the rudd pop-ulation then began to improve with the reduction in the size of the roach population. *L. intestinalis* also began to decline as the roach population was reduced. Based on numbers that I will not attempt to interject here, Clive surmised that *L. intestin*alis had a far greater impact on the roach popula-tion dynamics during the second cycle than in the first, and his reasoning seems justifiable.

Interestingly, the decline in roach population sizes suggested that a third cycle of change was about to begin, but the story stops at this point, with Clive's retirement from the academy. However, as Clive and I sat and talked that July afternoon in 2000, he told me of his investigation of events and changes in the Ley that might have occurred over the previous 100 years. He had already begun to examine old angling records and other such information that was available, including photographs of the lake from early in the twentieth century. Based on all of the information he had been able to assemble, he posited that the Ley had undergone several periods of change in the roach/rudd/pike community. Each of these changes resembled the two he had observed in the lake over the past 31 years, and the same one that appears to be occurring at the present time. As we talked, Clive remarked that "he had frequently wondered how many times history repeated itself in the Ley, and that it was tempting to speculate that parasites had a significant impact during each cycle." Another interesting facet of his long-term effort rests with the clear implication that, had he started his study in the middle 1970s, perhaps some of his conclusions regarding the impact of eutrophication and the nature of host–parasite interactions would have been quite different. They would have been even more so if the study had lasted for just three or four years. In other words, if some of these events and processes in the Ley, or any other body of water, are cyclic in any manner, then one must follow the cycle through completely before any solid conclusions can be drawn. The problem here is that not much can be concluded about the possibility of cyclic events in most ecosystems because long-term databases in most systems are nonexistent. Moreover, if long-term events are noncyclic, there would still be a serious question regarding reality in a system that is watched for just a few years. In other words, short-term field studies are like snap-shots of a garden. They give no hint about changes that will occur annually in that garden, recording only the state of the garden at a moment in time!

The work done by Clive Kennedy, his students, and other colleagues in Slapton Ley has been significant. In addition to the empirical findings they have generated, their efforts have helped to establish the clear advantage of long-term research in a single habitat. In these days, one of the greatest challenges facing a young academic is the securing of tenure. To accomplish this goal, a faculty member must publish papers. Unfortunately, one of the things that most academic deans can do quite well is count, i.e., publications. Long-term studies, in general, do not lend themselves to large numbers of papers, although in Clive's case this was not the case, and I can say that this was not the situation for our work in Charlie's Pond in North Carolina, or in Gull Lake in Michigan. Perhaps there is hope for the long-term approach after all.

There is another problem too. I was recently talking with my good friend Bob Overstreet, of the Gulf Coast Laboratory in Mississippi, and telling him of the research that one of my graduate students, Joel Fellis, was doing in Charlie's Pond. Joel was working on parasite communities in bluegill and green sunfishes and had, remarkably, recorded 16 species of helminth parasites in this small body of water. Bob asked me if we had data from 10 years ago. I had to say no. We have the data from 1983 for parasites in two species of pulmonate snails, but not for bluegill sunfishes. Why not? Because we were not looking ahead. We had no idea that our studies in Charlie's Pond would last 20 years. Who would have guessed it when Amy Crews did her first sampling back in 1982? Accordingly, I believe there should be some sort of repository for these sort of data sets. If there was, I believe there would be a natural incentive for investigators to go back periodically at various sites and examine their biological inventories over the long term. I think these sort of studies would be highly informative and provide important insights into the real dynamics of an ecosystem, not just a snapshot.

Clive Kennedy has given us an interesting perspective on what I consider a unique system and, in doing so, has pointed the way for others to follow. He played an important leadership role in the development of ecological parasitology on a worldwide basis, not only by creating a solid body of knowledge regarding Slapton Ley, but by developing innovative approaches to good questions. A personal highlight of my career has been the opportunity of knowing Clive and of having him as a scientific colleague, but, even more, as a gracious host whenever my wife and I have been invited to his home, and, finally, as a truly wonderful friend.

References

Esch, G. W., Kennedy, C. R., Bush, A. O., and Aho, J. M. (1988). Patterns in helminth communities in freshwater fishes in Great Britain: alternative strategies for colonization. *Parasitology*, **96**, 519–32.

Kennedy, C. R. (2001). Interspecific interactions between larval digeneans in the eyes of perch, *Perca fluviatilis*. *Parasitology* (suppl.), **122**, S13–21.

Kennedy, C. R., Shears, P. C., and Shears, J. A. (2002). Long-term dynamics of *Ligula intestinalis* and roach *Rutilus rutilus*: a study of three epizootic cycles over thirty-one years. *Parasitology*, **123**, 257–270.

3

Ecological studies in Charlie's Pond: "a stream of consciousness"

A professor can never better distinguish himself in his work than by encouraging a clever pupil, for the true discoverers are among them, as comets among the stars.

CARL LINNAEUS, QUOTED IN BIOGRAPHY OF LINNAEUS, BY BENJAMIN DAYDON JONES

In about 1968 or so, rumors began spreading locally about someone buying up large parcels of property in a rural area of North Carolina approximately 20 miles north-east of Winston-Salem where I teach at Wake Forest University. No one was quite certain who was purchasing the land, only that it was being bought. Then it was announced that Duke Power Company had acquired some several thousand acres for construction of a large reservoir to supply cooling water for two 1350-MW coal-fired steam generators.

My first direct experience with the new cooling reservoir, called Belews Lake, was in 1975, when I became Chairman of the Biology Department at Wake Forest. I had become involved in acquiring a long-term lease of about a hundred acres of land adjacent to the reservoir for the development of a departmental field station. Not long after, Ron Dimock, a colleague of mine in the Biology Department at Wake Forest, and I were successful in our effort to obtain funds from Duke Power to examine the population biology of the cestode *Proteocephalus ambloplitis*. This parasite is a potential pathogen for many centrarchid fishes, but primarily largemouth bass in the south-eastern parts of the USA. Of course, if there is the slightest chance of adversely affecting bass fishing in this part of the country, then money is readily available from any number of sources. In fact, I suspect there are more bass fishermen and bass boats in the south-eastern USA than there are would-be stock car drivers! This is why Duke

Power was willing to part with about $20000 to fund Ron and me for a couple of years. We were able to con – oops, persuade – Duke Power that providing us money to do some research on the potential problem would be a good investment, and we began working in the reservoir. As it turned out, it was a good investment, both for Duke Power and for us.

Small farm ponds are everywhere in North Carolina and whenever Ron and I went to Belews we would see a great many, never giving any of them much thought. Actually though, there was an exception, a pond that always intrigued me as we passed by. It is located immediately adjacent to the power station where a service road used to enter the plant site separates the pond from the main body of Belews Lake. The setting for the pond is exceptionally beautiful, especially in the fall when the towering maples and oaks that surround it on three sides begin their biannual metamorphosis (Fig. 15). The pond is small – less than 2 ha in surface area. It is fairly shallow and has some small stands of emergent vegetation scattered about, but not much. One of the things that always fascinated me about the pond was that it never seemed to fluctuate in depth as so many other small farm ponds in our area tend to do during the occasionally dry

Figure 15 Charlie's Pond, Stokes County, North Carolina, and the site of our long-term studies on the biology of *Halipegus occidualis* and other sundry parasites. As of 2003, in one way or another, research dealing with host–parasite interactions in this small North Carolina farm pond has contributed to the publication of 28 papers, reviews, or books. (Courtesy of Tim Goater.)

summer months. Moreover, it never seemed to accumulate any of the surface scum that many farm ponds are prone to when it gets into the "dog days" of August. We were to learn later that the pond is spring-fed, which accounted for its consistent water level and probably why it always seemed so pristine. We also discovered that under the dam and service road there is an outlet that drains water away from the pond and down into Belews Lake. The flow of water down and away from the pond explains why it was unaffected in the late 1970s by selenium pollution in Belews Lake, which I described in the Prologue.

The fish fauna in this little pond includes several species of centrarchids that are typical in the south-eastern USA, e.g., bluegills, green sunfish, and black crappie. We suspect the fish were trapped in the pond when the dam and service road were constructed. Most of the small farm impoundments in the south-east are referred to as bass/bluegill ponds, but oddly there are no largemouth bass present in this one, at least not at the present time. In the summer of 1987, however, a graduate student of mine who would contribute greatly to our early work there, Tim Goater, was catching green frogs in the pond one Saturday night. He heard a splashing noise in one of the coves and, out of curiosity, walked over to investigate. He was surprised to find a rather large fish flopping around in some very shallow water. Tim grabbed the fish with his bare hands and was surprised to find himself holding a 20-inch largemouth bass. In all of our collecting since then, we have not seen another largemouth bass and we have seen no evidence of bass spawning. We remain convinced that Tim caught the last of its kind in the pond.

The real story of our work in the pond began in the fall of 1982 when I taught a general parasitology course for a group of about 20 undergraduates. One of the students was Amy Crews, a most pleasant young woman who had started out at Emory University before transferring to Wake Forest the previous year. As I recall, she was rather shy, but became quite engaging as I came to know her during the semester. For some reason, she became hooked on parasitology and decided she wanted to go to graduate school and study parasites. I knew she would be successful wherever she went, but suggested she remain at Wake and do her Master's degree with me before going elsewhere for her Ph.D.

In the spring of 1983, after Amy was accepted into our graduate program, and while still a senior undergraduate student, she asked if she could get started with some preliminary research for her Master's degree. Anyone who has ever done ecology knows that to do a field study for the Master's degree is risky. It requires careful planning as there is but one seasonal cycle in which to complete the effort. If it isn't planned and executed properly the first time, then another year of fieldwork will be required and, at the Master's level, this is tough because of potential funding difficulties. So, I was agreeable to her suggestion and pleased with her enthusiasm in wishing to get an early start. A few days later, we drove out to our small pond next to Belews Lake and collected some snails, mostly *Helisoma anceps*, a common pulmonate in our region of the south-eastern USA. We brought them back to the lab and isolated them in small plastic jars filled with artificial pond water, the usual procedure. The next day, Amy began checking the jars for cercariae and found that several of the snails were infected. She came to my office, though, and asked me to come to the lab and look at something else that had appeared in her plastic jars overnight. I went down and had a look. At the bottom of several, there were large numbers of tiny, spherical things that I had never seen before. Moreover, they were not moving. We pipetted some from one of the jars and placed them on a glass slide so we could see them more clearly. After examining them with a compound microscope and checking the literature, we determined they were cystophorous cercariae (Fig. 16) of the

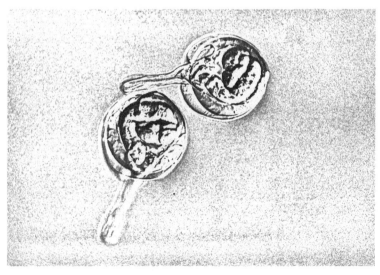

Figure 16 A pair of cystophorous cercariae of *Halipegus occidualis*. Protruding from the cyst is the "handle," which can be manipulated by the microcrustacean second intermediate host. Visible within the cyst is the cercarial body; just anterior to the cercarial body is the delivery tube through which the cercaria is shot explosively into the ostracod's hemocoel. (Courtesy of Tim Goater.)

hemiurid trematode, *Halipegus occidualis*. Although we did not know it at the time, Amy had found a parasite which, directly or indirectly, was to keep 15 more graduate students, a Canadian post-doc, and me, tied to this pond for nearly 20 years.

We learned that Wendell Krull had already worked out the complicated, four-host life cycle of the parasite back in 1935, but not much had been done with the parasite since. I emphasize again, Wendell Krull worked out the life cycle of *H. occidualis* in 1935. I reiterate this point because Jackie Fernandez and I were chided by none other than Miriam Rothschild in S. A. Ewing's *Wendell Krull: Trematodes and Naturalists* (2001) for omitting reference to this fact in our book, *A Functional Biology of Parasitism* (Esch and Fernandez, 1990), and I will not repeat that error of omission!

Because we knew then that Amy was going to be working in the pond for her Master's research, we figured that we had better give the pond a name. When I first went down to Par Pond and Savannah River Ecology Laboratory (SREL) to work with Whit Gibbons, I was greatly amused to find that he had given virtually every large cove in the reservoir a name, many times for a person, and occasionally for someone in his family. I liked the idea of personalizing these places, so I decided to call our little body of water Charlie's Pond, after my youngest son. Because we have worked this small pond for close to 20 years and published some 28 papers and monographs, based on research done there, anyone who has worked on the ecology of trematodes in snails, and frogs, knows about Charlie's Pond. For this reason, I decided to include the Charlie's Pond story in this collection of essays.

Before going any further, I need to introduce a few of the other characters who worked in the pond early in our research program (Fig. 17). Tim Goater was a Canadian student who came to Wake Forest in the same year that Amy started in Charlie's Pond. He had been an undergraduate at Brandon University where he had come under the flourishing influence of Al Bush, a well-respected ecological parasitologist who made his first research mark with John Holmes, working on the enteric helminth communities in lesser scaup ducks. Tim was to do his Master's research on the helminth communities of salamanders in the Blue Ridge Mountains of south-western North Carolina, and what a fine effort that turned out to be! Tim then stayed on for his Ph.D. to work in Charlie's Pond. Al Shostak, a Canadian National Science and Engineering Research Council post-doctoral fellow, came down to be with us at about the time Tim started. Al had done his dissertation with Terry Dick at the University of

Figure 17   Four of what my Canadian students fondly refer to as "Esch's lushes," loading gear into a departmental van at Charlie's Pond. They are (left to right) Shirley Shostak, Al Shostak, Tim Goater, and Julie Williams. (Photographer unknown.)

Manitoba on the ecology of the cestode, *Triaenophorus crassus*. Al was an excellent and well-trained ecologist/parasitologist. He was also a superb statistician who fitted well with the group that was beginning to assemble in the lab. Julie Williams arrived from Brent Nickol and the University of Nebraska to work on her Master's degree with me. Into our group also came Jacqueline Fernandez (Fig. 18) from the University of Santiago in Chile. Jackie had wanted to go with John Crites at Ohio State University, but John was nearing retirement and advised her to come to Wake Forest. She was already a highly experienced scientist, with something like 15 or 16 publications when she arrived. To this mix should be added Dave Marcogliese, another Canadian who, although he did not work in the pond for his dissertation, contributed a great deal to those who did.

After writing this essay, I had Tim Goater check it for accuracy. He made a very important observation, one that I thought should be included at this point. For some reason, I have been very lucky with the students who have come to do their graduate degrees with me. They have invariably been very bright and exceedingly hard-working, a combination of qualities that generally assures success. For the most part, they have also usually been very congenial and have interacted well with me, and with each other – another set of qualities that bodes well for success, in any

Figure 18 Occasionally, my old Chrysler LaBaron convertible would be used to haul people or equipment to Charlie's Pond. It was dubbed the Dean Machine, since I was Dean of the Graduate School at the time. Standing next to our special field vehicle is Jackie Fernandez. (Photographer unknown.)

laboratory. Something else occurs frequently if the chemistry is right. These students develop a "mutualistic camaraderie," as Tim called it. In doing this, they forge relationships with each other that will last a lifetime. You can see it whenever you go to one of the annual meetings of the American Society of Parasitologists. The Wake Forest "groupies" are together. It's not just true for students from Wake Forest, it's the case for any close-knit lab group, whether they were students at Nebraska with John Janovy, Jr. and Brent Nickol, or at New Mexico with Don Duszynski and Sam Loker, or at any number of other labs around North America where large numbers of students pass through a given graduate program. It is an interesting phenomenon, one that a sociologist ought to study some day.

Without question, one of the most unique features in the pond is the hemiurid trematode, *Halipegus occidualis*. When Amy began her work, we had no idea about how remarkable this parasite was, or that it was to become central to our subsequent "stream of consciousness" approach to research. By the time Amy finished, however, we did. Amy mostly focused on determining what trematodes were present in the pond, their seasonal dynamics, and then on deciphering their impact on the biology of the snail host, *Helisoma anceps*. She made several interesting discoveries regarding this snail–trematode system in the pond. First, she found that *H. occidualis* completely castrates the snail host and, second, the life span of the snail was found to be about 12 months long. Since the maximum life span of most of these snails is only 1 year, this means that the entire trematode community in the snails also turns over annually and, therefore, must be replaced each year. We have been surprised that nearly all of the eight parasites present when Amy started her work in 1983 were present the last time (1998) the snails were surveyed by Anna Schotthoefer, another of my Master's students. This suggests a remarkably stable ecosystem in which to work, again probably due in part to the spring-fed character of the pond.

I should emphasize another point regarding our work in Charlie's Pond. I have always been intrigued by long-term studies. By 1984, Tim Goater, Dave Marcogliese, Amy Crews, and I were finishing research on the population biology of *Crepidostomum cooperi*, an allocreadid trematode that uses centrarchid fishes as definitive hosts in Gull Lake, Michigan. I had started this work quite by accident some 20 years previously. One of my early Ph.D. students, Bob Morcock, had picked some mayfly subimagoes (*Hexagenia limbata*) off the screen of one of our lab windows at the W. K. Kellogg Biological Station (KBS), one warm night in August in 1966. These mayflies had metacercariae of *C. cooperi* under their abdominal skin, and their darkly pigmented cyst walls made them easy to spot, and count. Just for the fun of it, we began collecting about 200 of these mayflies soon after their emergence in early August of each year. After I stopped going to KBS in 1974, an old friend, Art Weist, would make the collections for us each summer, which we continued through 1984 when we finally stopped. The theme of the mayfly/*C. cooperi* story is again related to the stream of consciousness idea. Initially, we had no clear reason for collecting the mayflies and counting their parasites, except to see what happened from one year to the next. A few years into the study, however, we learned that Gull Lake was undergoing eutrophication. On reading a really good paper on the biology of *H. limbata*, we realized that we were on to something potentially significant and we then pursued the work with real vigor. This study personally taught me the value of long-term parasitology research, something that has been lacking, except for a few studies by Clive Kennedy in Slapton Ley in Devon, England, John Aho and Joe Camp in Par Pond, in Aiken, South Carolina, and H. D. Smith in Babine Lake, British Columbia. Whereas we did not begin the Charlie's Pond work with a long-term effort in mind, it ended up being just that, a long-term research study. I would urge any young parasitologist just starting out professionally to find some place nearby and begin following the system on an annual basis. It does not require a huge amount of effort, simply persistence, patience, and an open mind. Like us, you may be truly surprised by what you turn up.

Despite the long-term character of the work done in Charlie's Pond, as I said previously, it is *H. occidualis* that makes this pond so fascinating. Most hemiurids are stomach flukes of marine fishes, although there are several species of *Halipegus* infecting ranid frogs in North America and Europe. There is even a new species in Costa Rica, *Halipegus eschi*, named for yours truly, which was just described by one of my former graduate students, Derek Zelmer, together with a long-time friend at the University of Toronto, Dan Brooks. I have seen a specimen of this new parasite, and it is impressive, as far as flukes go! Imagine having a parasite named for you, a beautiful fluke whose congeners would be the focus of your research and that of your students for nearly 20 years. It was indeed quite an honor. Adult *H. occidualis* occur under the tongue in the mouth of the green frog, *Rana clamitans*. Living under the tongue means they can be easily counted and, most importantly, without killing the frog! All that is necessary is to catch a frog, open its mouth, lift its tongue, and count the parasites. More-over, immature parasites can be distinguished from mature ones by their size and color. If the frog is toe-clipped and returned to its site of capture, it can be recaptured at a later date and the parasites can be counted again. And, this can be done over and over. To my knowledge, this species, and *Halipegus eccentricus*, a hemiurid that occurs in the Eustachian tubes of the green frog, are the only endoparasitic trematodes whose adult stages can be enumerated without killing their definitive host. This truly remarkable characteristic meant that, for the first time, the infrapopulation dynamics of an adult helminth could be followed with great precision, in the same host, over time. Previous to this, the best that one could do without us-ing the "kill-and-count" method was to enumerate the eggs of helminth parasites in the stools of their hosts and then estimate parasite population sizes based on egg production per worm – a highly imprecise method for quantifying worm numbers.

Adults of *H. occidualis* release eggs which are shed in the frog's feces, but they do not hatch, as is the case for a number of fluke species. The eggs must be consumed by *Helisoma anceps* where they then hatch in the snail's gut. The larval stage (miracidium) emerging from the egg migrates first to the snail's hepatopancreas where it transforms into a sporocyst, which, in turn, give rise to rediae that eventually spill over into the gonads. Rediae possess a mouth and a primitive gut, which they can use directly to ingest host tissue. When examined microscopically, the rediae of *H. anceps* are quite distinctive. In a rather disgusting way, they remind me of fly maggots.

Eventually, cercariae of the cystophorous type are produced by the rediae and released from the snail. They are most interesting, for a number of reasons. Unlike most cercariae, they do not swim, remaining motionless on being shed. The next intermediate host, usually a benthic-dwelling ostracod (a common, freshwater microcrustacean), must eat them. When they are picked up by the ostracod, the second intermediate host, their surface will be punctured by the ostracod's mouthparts. Inside the outer covering of the parasite, there is a sudden change in the osmotic pressure, causing a relatively long delivery tube that had been coiled inside to evert explosively. When the delivery tube emerges, it immediately penetrates the gut wall and the body of the cercaria is shot through the tube like a cannon ball, directly into the ostracod's hemocoel! Once inside the ostracod, the parasite immediately begins to absorb water and is gradually activated from its anhydrobiotic state. Writing in *Novitates Zoologicae*, Miriam Rothschild (1938) commented on the evolution of this mechanism. She said, "The extraordinary degree of specialization shown by this group of cercariae [the cystophorous types] is unique, and it is difficult to conceive how the delivery apparatus, with its peculiar function in the life history of the cercariae, could have arisen." She continued, even though "this [encystment in the tail] was probably the first stage in the evolution of the present peculiar method of excystment of the cystophorous cercariae, it gives us no hint as to how the extremely complicated and delicately adjusted delivery system came into being."

Another of my graduate students, Derek Zelmer (Fig. 19), continued with the life-cycle research begun by Tim Goater. He found that, within the ostracod, the metacercaria will increase in size and, after about 5 weeks, will become infective for the green frog, at least under laboratory conditions. In the field, however, we doubt the parasite reaches full development in the ostracod, which is consumed by the next host in the life cycle, a dragonfly nymph, before this can happen. When the nymph eats the ostracod, the metacercaria remains in the lumen of the nymph's gut. The dragonfly nymph is thus a paratenic host, one that is required to bridge a trophic, or feeding, gap in the parasite's life cycle. We are confident that it is quite unlikely the parasite is transferred directly

Figure 19 Derek Zelmer spent a great deal of time wading in Charlie's Pond as part of his Ph.D. research on the population dynamics of *Halipegus occidualis* in green frogs. Derek, as well as two other of my Ph.D. students, Tim Goater and Eric Wetzel, took advantage of the unique opportunity of being able to monitor recruitment of a macroparasite without having to use the traditional "kill-and-count" method of censusing. (Photographer unknown.)

to the frog via the ostracod in the field. In fact, I seriously doubt that a ranid frog can see these microcrustaceans, let alone eat one, except accidentally. A dragonfly that has just emerged from a pond and is sitting on the stem of a *Typha* plant as it undergoes metamorphosis – now that is another matter. This is a food morsel that would be quickly sought by any insectivorous ambush predator, e.g., a hungry green frog, and is most likely how our parasite reaches its definitive host in Charlie's Pond.

Rather than providing details of the empirical data generated by Tim, Julie, Jackie (Fig. 20), Al, and the others who followed, I will describe only in more general terms what they found and how they accomplished it. Tim had then, and still has, a great fascination for amphibians. Because of this interest, he was to place most of his attention on the population biology of the adult parasite in the frog hosts. His collecting methods, I have long felt, were exceptionally novel. Tim loved to sail and was the proud owner of a windsurfer – a flat, one-person sailing board. His collections in Charlie's Pond were made at night. He would tow the windsurfer into the

Figure 20   Jackie Fernandez, me, Julie Williams, and Tim Goater, taking a break from work in Charlie's Pond. (Courtesy of Tim Goater.)

pond and load all of his collecting gear on the tiny platform. Then, with a miner's lamp on his head, he would enter the pond, climb aboard his windsurfer, and hand-paddle the tiny craft around the edge of the pond. When he located a frog (Fig. 21), he would capture it by hand, open its mouth, lift the tongue, and count the parasites. After taking appropriate measurements of the frog and making an identifying toe clip, he returned the frog to the exact place at which it was captured. He continued around the entire pond, capturing frogs and recording his field data as he went, using his windsurfer as a laboratory bench. Tim is an interesting person to be around, and he was really excited about his work. I will never forget the morning I showed up at work and he came charging into my office with fire in his eyes, exclaiming he had recaptured a frog that he had marked 3 years previously and had not seen since!

Tim was also interested in the population genetics of the parasites, so he would remove one or two with a pair of forceps and place them into a vial on dry ice that he also carried on his windsurfer as he collected. These frozen parasites were then shipped to Peg Mulvey, a first-class population geneticist at SREL who would do starch-gel electrophoresis for us. About 6 weeks after one of his shipments, Tim received a call from Peg saying that four of the parasites he had sent down were clearly genetically different

Figure 21   Tim Goater, wearing his miner's lamp, sitting on his windsurfer, while "frogging" one night in Charlie's Pond. This was Tim's last night of frog sampling for his long-term dissertation study of *Halipegus occidualis*. (And, yes, that is a beer sitting on Tim's unique "lab bench," celebration time for a field parasitologist!) The frog he is holding is one he had captured 3 years previously and had not seen since. (Courtesy of Travis Knowles.)

than any of the others and, "what was going on?" Tim went to his field notes and found that these four flukes had been removed from the eustachian tubes, not from under the frog's mouth. On further investigation, he determined these parasites were new to his frogs, and that they were *Halipegus eccentricus*, another hemiurid fluke. The literature indicated that *H. eccentricus* required yet another pulmonate snail, *Physa gyrina*, as the first intermediate host. Sure enough, on examination of the cercariae being shed from *P. gyrina*, he discovered they were clearly distinct from those being shed from *H. anceps*. This species of snail was not present when Amy began working in the pond in 1982, but Tim found it in 1986 when he went looking for it. We believe the new snail and *H. eccentricus* had colonized the pond, unbeknownst to us, at some point during those first 3 years of our work. As we were to discover later, the new snail also provided the means for colonization by several other new trematodes into the pond, as well as *H. eccentricus*. This chance observation was to lead us into new opportunities for research in the pond because we could now do a comparison of the two congeners that were obviously similar, yet strikingly different in a number of interesting ways. We had entered another "stream of consciousness" phase of our research in Charlie's Pond.

As students frequently do when they are intensively working on similar projects, they talk about their research and come up with new ideas and different approaches to problems. During one of these "bull sessions," Al Shostak, Julie Williams, and Tim decided they wanted to know if the stress of infection influenced the success of infected snails in making it through the winter months, and if any overwintering snails lose their infections? To answer these sorts of questions, however, they would need to find a way to mark some infected snails, return them to a specific site in the pond, and then recover them the following spring. To be successful, they also needed a snail that was not very vagile. Fortunately for them, *H. anceps* was an ideal species to use in such a study. They collected approximately 500 snails (Fig. 22), determined the nature of infection in each one, and then reintroduced them at a specific site in the pond in the late fall. Before they returned them to the pond, they developed a system for marking the snails, a very simple method. They covered a portion of a snail's shell with enamel paint and, after it dried, placed an identification number written in indelible ink over the enamel paint. Hand-held calipers

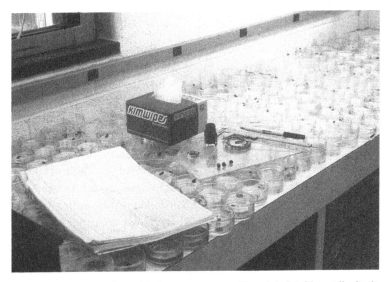

Figure 22   Large numbers of *Helisoma anceps* were collected, isolated in small, plastic jars, examined for released cercariae, then marked with enamel paint and numbered with indelible ink before being returned to the pond during the mark–release–recapture program. (Courtesy of Jackie Fernandez.)

Figure 23   In addition to marking snails, the shell height of each snail would be measured using hand-held calipers so that the growth rates of recaptured snails could be measured. (Courtesy of Jackie Fernandez.)

were used to measure all marked snails so that their growth rates could be determined on recapture (Fig. 23). In the early spring, nearly 20% of the marked snails were recovered. They found that 74% of the infected snails had retained their infections but, in the other 26%, the parasites had been lost. Moreover, they discovered that several of these snails that also had been previously castrated by *H. occidualis* and had lost their infections over the winter were capable of regenerating their gonads and were actually producing eggs. When they found the first previously infected snail producing eggs, they were initially surprised and each blamed the other for recording a false-positive infection the previous fall. They were relieved, however, when they observed that several other previously infected snails were shedding eggs in the spring and, on dissection, that their gonads were intact. In effect, what they found was that snails were not permanently castrated, even though infected snails stopped shedding eggs. It was a good lesson for them, and me, because it clearly emphasized the notion of dynamic and complex interactions involving snail–trematode systems.

Subsequently, Brian Keas undertook a very complicated laboratory investigation in which he tested the effects of diet and infection by

H. occidualis on the growth rates and fecundity of *H. anceps*. Among other things, Brian found that fecundity in the field was significantly lower as compared with snails in the lab fed on a high-quality protein diet. However, he was also able to demonstrate that low-quality lab diets reduced fecundity to significantly lower levels than those observed by Amy Crews in her earlier studies in Charlie's Pond. This suggested to Brian that, whereas *H. anceps* in Charlie's Pond were not deprived of protein, their diet was affecting fecundity by reducing it from levels seen with a high-quality diet in the lab. Tim, Al, and Julie had observed the loss of infections in overwintering (hibernating) snails using their innovative mark–release–recapture procedure. This observation was replicated in Brian's study, except that spontaneous loss of infection was induced by low-quality diet in the laboratory. These findings were salient because they strongly suggested that certain energetic phenomena were vital to successful population biology of both snails and their trematode parasites, reinforcing the idea of dynamic and complex interactions involving snail–trematode systems.

It was about this time that Tim and Jackie Fernandez began working on another problem in the pond, one that was to yield some very interesting information on another aspect of our snail–trematode system. Those who work with snails have probably observed a tiny, but still obvious, oligochaete annelid that lives as a commensal in the mantle cavity. This organism, *Chaetogaster limnaei limnaei*, is known to feed upon a range of microorganisms that may enter the cavity, including miracidia of digenetic trematodes attempting to infect the snail. There has even been speculation that the oligochaete may offer protection to the snail against invasion by trematode miracidia. When Tim and Jackie were doing some preliminary work on *H. anceps*, they noticed that *Ch. l. limnaei* were consuming cercariae of *H. occidualis*. They wondered if the trematode might be serving as a food resource for the oligochaetes. Optimal foraging theory would suggest that our parasite may be contributing to predator density, especially since the cystophorous cercariae of *H. occidualis* do not swim on being shed and prey vulnerability is thought to be an important mechanism for selective predation in nature. With this in mind, they undertook a comprehensive study to test their hypothesis regarding the population biology of the oligochaete. Whereas they found that the annual cycle of reproduction and mortality within the snail population influenced the population biology of *Ch. l. limnaei*, they also reported significantly higher oligochaete densities in infected snails than uninfected ones. These data thus supported the notion that there should be a positive correlation between predator and prey densities and that a "dominant predator response," according to Sih (1984), should occur when immobile prey are involved. I was particularly pleased with the outcome of this study because I distinctly recall a well-respected colleague commenting at the time, "Why are they doing this work? Everyone knows that *Chaetogaster* is just a symbiont in the mantle cavity of snails." Jackie and Tim had taken a simple trematode–snail system, however, and used a third, supposedly innocuous organism cleverly to support some fundamental ideas regarding predator–prey theory.

Early in our research in Charlie's Pond, I gave serious thought to the idea of population genetics in relation to the biology of snail–trematode interactions. I suppose that many investigators wonder at some point why some hosts are infected and not others. Is it by chance, or is there some sort of predilection to infection? If the former, is it simply a matter of being in the wrong place at the right time, or is another, more complicated, factor involved? If the latter, can we equate predilection with predisposition? Is the genetic factor related to host behavior in some way, or is it something more complicated such as host immunity? Could it have to do with virulence factors associated with the parasite in some manner? I could continue asking these questions, but the genetics of host–parasite interactions are far more complex than I can explain here, assuming I could explain the nature of these interactions under any circumstance, because I cannot. I raise these issues because they are questions that ultimately must be answered.

As I just alluded to, we thought the question of whether snails infected with *H. occidualis* are genetically different from uninfected snails could be simply answered with some starch gel electrophoresis. We were not looking for anything in particular, just for an indication that might point us in a direction. We were surprised when Peg Mulvey, our friend at SREL, reported that she had found that two enzymes were different in infected and uninfected *H. anceps*. During the review process in publishing the paper dealing with these genetic differences, however, I can clearly recall one of the referees raising the issue of bias in our sampling procedure. This greatly annoyed me because we had used the standard collecting protocol of "pick up snail and toss in bucket," the one generally employed by anyone who has spent time doing any sort of basic survey. So, Julie Williams, as part of her Master's research in Charlie's Pond, made a concerted effort to test the hypothesis regarding the implied collecting bias. She cordoned off

Figure 24   Julie Williams, using her "snail Braille" technique to collect *Helisoma anceps* in Charlie's Pond. This study was one of the clearest demonstrations of the importance of microhabitat partitioning in the transmission of digenetic trematodes. (Courtesy of Tim Goater.)

about 45 m of shoreline in one small cove and marked each 1-m stretch of shoreline with a stake and a number. Then, using a table of random numbers to select specific quadrants during each sampling period, she set out to collect as many *H. anceps* as she could find. She identified the precise location of each snail removed from the pond, recording the depth at which the snail was captured, its distance from shore, and the type of substratum on which the snail was sitting when taken from the pond. The procedure required her to do extensive groping in turbid water where the bottom of the pond could not be seen. She appropriately dubbed her technique "snail Braille" (Fig. 24).

We were both flabbergasted when she found that, indeed, we had been using a biased sampling protocol and the referee was, at least in part, correct. The bias was not, however, related to genetic differences between infected and uninfected snails. Julie instead proposed that it most likely had to do with the way in which the parasite was recruited by the snail, in combination with the behavior of the green frog definitive host. She found that the probability of finding an infected snail was directly correlated with the depth of the water. To wit, the shallower the water and the nearer to shore, the more likely she was to pick up a snail infected with *H. occidualis*. The explanation for this observation is basic to the way in which these frogs forage. The green frog is an ambush predator and sits in shallow water, waiting for its prey. Presumably, although we do not know for certain, it defecates as it sits. We know that the parasite's eggs are shed in the frog's feces and then accidentally ingested by the snail as it grazes. Moreover, the mark–release–recapture studies of both Jackie and Julie indicated that the snail host, *H. anceps*, is not vagile, so that those infected in shallow water tend to stay there. In typical surveys, snails are simply tossed into a pail or bucket as they are collected, disregarding any consideration of site segregation with respect to where they are collected, e.g., in shallow or deep water, on leaf litter, or a log, or the pond's substratum. The results of Julie's efforts suggested to us the notion of spatial heterogeneity for the transmission of *H. occidualis* in Charlie's Pond. This idea was to be subsequently incorporated into the concept of landscape ecology for our parasite in the context of another study that was to follow Julie's a few years later.

In the middle to late 1980s, Armand Kuris at the University of California at Santa Barbara and Wayne Sousa at the University of California at Berkeley published several very interesting papers on the community ecology of trematodes in snail hosts. Both are really excellent scientists, as is Kevin Lafferty who was subsequently to co-author several papers with Armand, dealing with the same issues. However, a number of folks who work on snail–trematode systems, including my students and me, do not entirely agree with the message they were propounding. Let me explain the nature of the disagreement. Examination of the snail–trematode literature shows that, in the overwhelming majority of studies, the prevalences of trematodes in snails are relatively low, e.g., in the range of 3–8%. In addition, most snails are infected with a single species of trematode, although there are some exceptions to both of these generalizations. Low prevalences in snails depend on a range of environmental and transmission factors, and this assertion is not disputed in any way. In fact, with a few notable exceptions, success in transmission is usually not very high for most eukaryotic parasites, which is one reason for their very high fecundity. Explaining the dearth of multiple-species infections in snails is another matter. Will Cort, Don McMullen, and Sterling Brackett were the first to conduct an extensive quantitative study on trematodes found in snails in several northern Michigan lakes back in 1937. They found the trematode infracommunities to be quite depauperate. Indeed, in all of the hundreds of thousands of snails examined around the world since 1937, the most species recovered from a single snail has been four, and this has been recorded only three times. Why the depauperate infracommunities? According to the Kuris–Sousa–Lafferty perspective, the explanation rests mainly with the idea of negative interaction, more precisely with predatory rediae. As mentioned earlier, rediae possess a mouth, a muscular pharynx, and a primitive gut, and are known to consume host tissue by tearing and swallowing. Apparently, some are also capable of directly consuming larvae of established subordinate trematode species, or of capturing and eating newly invading larvae, which will then result in single-species infections of the species with the dominant rediae. Kuris and Lafferty have even gone so far as to create a complex system of dominance hierarchies, based largely on rediae size. They say, for example, that trematodes with rediae in their life cycles will dominate those species without rediae, and that those species with large rediae are dominant over species with small rediae. Much of their conclusions are based on work done with the marine snail, *Cerithidea californica*, found on the coast of California.

In contrast, Larry Curtis, working with the mud-flat snail, *Ilyanassa obsoleta*, along the coast of Delaware, has challenged their conclusions regarding the role of predatory rediae in structuring trematode infracommunities, opting instead for an explanation that rests on the idea of temporal variation in parasite recruitment and snail age. Our results in Charlie's Pond tend to support those of Curtis. Jackie Fernandez, for example, examined 4899 *H. anceps*, using our mark–release–recapture procedure, and found 1485 to be infected. However, she found only eight double infections. Based on these findings, one might infer this as evidence for competition/predation in her system, but she concluded otherwise. She determined the depauperate infracommunities were more like related to the temporal visitation of the definitive hosts to the pond, in combination with the age of snails at the time parasite eggs are released into the pond. Some snails, for example, are refractory to infection by a given parasite if the snails are not the appropriate size. Other factors affecting infracommunity organization in Jackie's study included differential mortality of snails infected with certain trematode species, constant recruitment of one trematode species over time, and the consistent timing of the snail population's mortality and replacement with a new cohort during the same 6–8-week period each year in Charlie's Pond.

Immediately after Jackie completed her work on snail trematode communities in *H. anceps*, Scott Snyder from the University of Nebraska was to join our crew as a graduate student working on his Master's degree. Continuing the "stream of consciousness" approach to our work in Charlie's Pond, he immediately set about doing basically the same sort of research that Jackie had done on *H. anceps*, except his was focused on the new colonizer to the pond, *Physa gyrina*. Scott's results were absolutely striking in comparison to those generated by Jackie Fernandez. From the beginning of our studies in the pond, *H. anceps* has consistently had seven species of flukes which cycled through it, and one more which showed up somewhat erratically from year to year. Three of these species are acquired when the snails eat the parasites' eggs, and four are recruited via penetration of miracidia. In *P. gyrina*, Scott found six species of trematodes, three of which are acquired by ingestion of eggs and three via miracidia penetration, not unlike the disposition of parasites in *H. anceps* with respect to the mode of snail infection. In other words, the mechanisms of infection by the flukes in the two snail species are quite similar. Moreover, two of the trematode species use both snails as their intermediate hosts and two others are congeners, demonstrating even more similarity in the trematode infracommunities within the two snails. Because of these matching patterns, we also expected to see much the same in the number of multiple infections occurring in the two sympatric snail species. Recall that Jackie observed eight double infections among the nearly 2000 infected *H. anceps* she recovered from Charlie's Pond. In contrast, however, Scott found that nearly 20% of 467 infected *P. gyrina* in the pond were infected with more than one species of fluke.

Scott maintained snails with multiple infections in the lab to establish any dominance patterns. If a snail with multiple infections of the same parasites consistently lost the same parasite first, he could be certain of dominance, but this was not the case. Indeed, there was no pattern at all. He concluded that dominance was not a factor in determining trematode infracommunity structure in *P. gyrina*, just as Jackie had established an absence of dominance in *H. anceps,* but for totally different reasons. Jackie relied on a variety of temporal and other factors to explain the lack of dominance, whereas Scott contended that trematodes in *P. gyrina* simply coexisted, again with no dominance being apparent. I first met Kelli Sapp when she participated in our department's National Science Foundation (NSF)-sponsored summer program for undergraduates. She went on to work in my lab on her Master's degree, conducting some interesting research at the pond by using infected and uninfected sentinel snails to examine a spatial component of recruitment by *P. gyrina* and *H. anceps*. The results of her study strongly supported the conclusions reached by both Jackie and Scott.

Based on all of these efforts, we have concluded that, whereas dominance may be important in determining structure in some host species, many other factors are involved to the exclusion of negative interaction via direct competition or predation, certainly the case in Charlie's Pond. Why are *I. obsoleta* and *C. californica* so different, especially since both are prosobranch snails, they occupy similar habitats, and they must be exposed to a slate of trematodes that are at least somewhat similar? The same question could be asked of *P. gyrina* and *H. anceps* and they are in the same pond, along with several of the same parasites or their congeners as part of their respective trematode infracommunities. I cannot offer a specific resolution to this question because I do not believe there is a single answer. In other words, I concur with Kuris, Lafferty, and Sousa that interactive forces may affect infracommunity structure when certain trematode species combinations occur. However, the bottom line suggests that a mixture of host age, time, space, individual susceptibility/resistance, as well as various interactive factors, are collectively involved, and that one of these factors is not predominant to the exclusion of any other.

As I noted earlier, Tim Goater was the first to deal directly with adult *H. occidualis* in the pond's green frogs. After Tim, however, two other graduate students spent a great deal of time in following up on his work. The first of these was Eric Wetzel. Eric came to Wake Forest to work on his Master's degree with Pete Weigl, a recognized expert on flying squirrel ecology in the south-eastern USA. On completing his thesis with Pete, he then moved to my lab. We decided that he needed to extend some of Tim's observations on *H. occidualis*, but to include *H. eccentricus* for comparative purposes. He was also to undertake what was to become an interesting study of the population biology of *H. occidualis* metacercariae in the dragonfly nymphs in Charlie's Pond. When he began his dissertation research, we were under the impression that the life cycle of *H. eccentricus* involved *P. gyrina*, ostracods, and then *R. clamitans* tadpoles. The literature suggested that tadpoles supposedly ingested infected ostracods and that metacercariae made their way into the tadpoles' stomachs where they remained until metamorphosis of the tadpoles. Then, the parasite was said to migrate from the stomach up into frog eustachian tubes where they matured sexually. Eric's work, however, demonstrated this could not be the case. First, on necropsy of approximately 100 tadpoles, no metacercariae were found. Second, his mark–release–recapture of adult green frogs revealed the presence of sexually immature *H. eccentricus* in second-year frogs. Since sexual maturation of the parasite is rapid, there is no way that immature parasites would be present in second-year frogs. Whereas *H. occidualis* is quite amenable to laboratory and field manipulation, neither Eric nor any of my other students have been able to complete the life cycle of *H. eccentricus* in the lab; indeed, we were never able to infect *P. gyrina* successfully in the lab either.

Eric also made a series of observations on the way in which adult *H. occidualis* infrapopulations seemed to build up under the tongues of the frogs, then suddenly disappear. He hypothesized that the parasites were irritating the tender mucous membranes under the tongues, ultimately causing the tissues to slough off and be swallowed, along with the adult parasites that might be attached to these sloughed tissues. Histological work lent credence to this line of thinking. The mark–release–recapture studies on frogs revealed a remarkable dynamic in parasite infrapopulation sizes, with sharp increases and declines in relatively short periods of time. The sloughing of irritated host tissue and the associated parasite infrapopulation accounted for the sharp decline in parasite densities. He conjectured that this is a form of density-dependent regulation of the parasite's infrapopulation size, an idea first suggested by Tim Goater several years earlier. Predation upon heavily infected dragonflies accounted for the rapid increases in infrapopulation size.

Based on Eric's findings, we can safely conclude that the dynamics of parasite infrapopulation sizes in green frogs are quite radical, capable of changing significantly in short periods of time. The only way of really knowing what is occurring in a parasite infrapopulation over time, however, is by following these changes in individual hosts through a comprehensive mark–release–recapture program. Unfortunately, except for some monongenes and a few other other ectoparasites, no other parasitic helminth systems lend themselves to this approach.

Derek Zelmer did his Master's work on perch parasite communities in several large reservoirs in the Canadian Province of Alberta. He received his degree at the University of Calgary where he studied with Professor Hisao Arai before coming down to work in my lab. When Derek arrived at Wake Forest, we initially planned for him to do an ecology dissertation involving helminth parasites in stream fishes, but this was to change soon after he came. The fall of his arrival was a sabbatical semester for me and part of what I did was to travel extensively in the UK and Ireland, visiting with various parasitology colleagues along the way. I remember sitting and talking with Clive Kennedy in his office at the University of Exeter when an idea suddenly popped into my head, one that I thought had some real dissertation possibilities. Since I had no other doctoral students who were just beginning, I decided that Derek should be the one to tackle the project, but I had to talk with one more person about it before I could be sure it would work. This person was Celia Holland at Trinity College in Dublin. As fortune would have it, I was scheduled to visit with Celia as part of our Ireland excursion.

By the time we reached Dublin, the idea had percolated in my mind long enough that it had begun to take reasonable shape. The plan was really quite simple and was prompted, in part, by a manuscript Celia had submitted to the *Journal of Parasitology*, of which I was Editor at the time. The paper had to do with the comparative costs of treating human patients with enteric helminth parasites. A little insight is necessary to see how treating a disease like ascariasis in Africa can be related to trematode parasites in the mouths of green frogs in North Carolina. As I mentioned in the Prologue, in 1971 Harry Crofton published his seminal papers on parasite population ecology and epidemiology. Part of what he said was that many helminth parasites are aggregated, or contagiously distributed, in their hosts. That is to say, for many species of helminths, most of the parasites within a given host population are present within just a few individuals. Accordingly, most of the members of a given host population are not infected, or carry relatively few parasites of a given species. This is just as true for *Ascaris lumbricoides* in an African village as it is for *H. occidualis* in the green frogs in Charlie's Pond. Crofton's idea eventually led several epidemiologists to develop several experimental protocols for the treatment of ascariasis and other similar diseases in humans. For example, with mass treatment, everyone in a given village would receive an anthelmintic drug, irrespective of whether they were infected or not. This is an expensive approach in most Third World countries where there is little money to pay for drugs, or for their delivery, and also because not everyone in the village has the parasite and, therefore, requires treatment. In selective treatment, therapy is provided only to those who are heavily infected. This is also expensive since the stools of everyone in the village must be checked for eggs to determine who is heavily infected and who is not. Targeted treatment involves therapy for groups who are suspected of being susceptible within a host population as predefined by age, sex, or any behavior that might produce a predilection for infection. In large measure, the idea behind these latter two approaches rests with the notion that, if the parasite's population size can be driven below some threshold, then transmission will be disrupted and the parasite will gradually disappear. There are problems with these two protocols, however, ranging from ethical considerations in the use of human volunteers to the use of indirect methods (eggs in the stool) for estimating prevalence and abundance of the parasite in the treatment group. The testing of selective or targeted treatment protocols in humans, therefore, is a dicey proposition.

In contrast, however, we had a system in which the parasite populations could be easily manipulated without killing the host and with the certainty that we could eliminate whatever number of parasites we desired from a given frog host in the pond. Moreover, we did not have to deal with the vagaries of human behavior, or with problems related to securing permission, or ethical concerns. We had a system where we could, therefore, easily test the selective treatment protocol. It seemed like a good approach and when I arrived for lunch with Celia at the Trinity College faculty club, I sprung it on her to see what she thought. Celia thought it was a grand idea. So, after a great lunch, I e-mailed Derek from her office, telling him to put his stream project on hold. I was coming home to try a new idea out on him regarding his dissertation research. When I returned and told him about what I had discussed with Celia, he bought it, and immediately developed a plan of execution to test the selective treatment protocol using the green frogs and *H. occidualis* in Charlie's Pond.

The first summer, he initiated another mark–release–recapture program in the pond, counting the parasites as he obtained the frogs. With these data, he was not only able to obtain an estimate of the total numbers of parasites in Charlie's Pond, he was also able to estimate the total number of frogs present as well. Then, halfway through the second summer of his collecting efforts, he removed 51% of the adult *H. occidualis* from the pond. He added one twist to his study. He left one parasite in the mouth of each frog from which he had taken the 51%. He did this so that the overall prevalence of the parasites in the green frogs was not changed. The next year, he monitored the extent of aggregation, parasite densities, and trends in colonization. All three returned to pretreatment levels, indicating that parasite removal had no effect on the parasite's population biology and that selective treatment, in his system at least, had no effect. He argued that, because of the focal nature of transmission dynamics from the frog to the snail, e.g., ingestion of eggs by the snail rather than miracidia penetration, regulation was governed by prevalence of the parasite rather than by density. He extrapolated his conclusions to some of the enteric helminths of humans that are likewise transmitted via eggs, asserting that repeated selective treatment might affect host morbidity, but would not influence long-term parasite transmission. This was a bold contention, one that certainly requires verification through well-controlled field research. A problem with several of the current epidemiological approaches to some of the enteric helminth diseases in humans is that they are largely based on predictions generated by mathematical models, which do not always rely upon accurate empirical databases. Finally, one must always be aware that a mathematical model is no more than a sophisticated, usually mathematical hypothesis and that, as such, it must be subjected to the rigors of careful experimentation and testing. Despite extensive fieldwork on many of these parasites for the last hundred years, solutions for their elimination remain problematic, but more about this situation in Chapter 7.

There was still another theme that was to emerge, especially based on the work of Tim Goater, Eric Wetzel, and Derek Zelmer in Charlie's Pond, and that had to do with spatial factors in the transmission of *H. occidualis*. One day in the spring of 1996, before Eric left for his job at Wabash College up in Indiana, and while Derek was still pursuing his selective treatment study, I asked the two of them if there was any evidence of spatial transmission foci in the pond. I was reminded of Julie's observations regarding the location of infected snails in shallow water and wondered if something like this might apply to the infection process in green frogs. Using information generated by him and Eric, Derek began to assemble additional data with which to examine the question of parasite transmission and spatial factors. When the distribution of infected frogs was plotted on a detailed map of the pond he found that, indeed, there was clear spatial heterogeneity, with four transmission "hot spots." He then carefully examined these four sites to see what they had in common and found that they were all shallow, fitting nicely with Julie Williams' representations regarding transmission of *H. occidualis* eggs from frogs to snails. Moreover, each site had a stand of *Typha* spp., i.e., emergent vegetation, and it was surmised that this was key to the parasite's spatial heterogeneity, for three reasons. First, the emergent vegetation provided cover for the frogs from potential predators. Second, with emergent vegetation, dragonfly nymphs had a place on which metamorphosis could occur above the water line, and gave opportunity for an ambush predator to see its prey, making these nymphs vulnerable to the green frog. Third, the vegetation also served as a trap within which leaf litter could accumulate. An abundant supply of leaf litter presented an excellent nutrient resource for grazing by *H. anceps*, the first intermediate host for *H. occidualis*.

As I reflect on it now, the work in our pond not only emphasizes the concept of landscape epidemiology, it provides insight with respect to the idea of both spatial and temporal scales in the transmission of parasitic organisms. Both Wincenty Wisniewski and E. N. Pavlowski emphasized these ideas more than 50 years ago. We have learned a great deal about the rapidity with which parasite populations and communities can change and wonder if this dynamic quality is not a feature of more host–parasite systems than just ours in Charlie's Pond. We were also working in the pond when an important colonization occurred, i.e., the sudden appearance of *H. eccentricus*. I should note here that this parasite has now apparently disappeared from the pond, just as innocuously as it had appeared. I indicated at the outset of this essay that our work in Charlie's Pond reflects what I call a "stream of consciousness" approach to research. Some might assert that this path is like saying, "if you do not know where you are going, any road will get you there." This is a valid criticism, I suppose. However, unless one is constrained by the limits of a federal or state grant, or by the requirements of a narrow-minded dean or departmental chair for many publications, I would suggest that this is an excellent way of doing science. It sure as hell allows one an exceptional level of freedom to follow a trail of discovery and, having been constrained myself by federal guidelines imposed by granting agencies, I am quite certain it is the most fun.

As I mentioned earlier, one other contribution made by the investigations that focused on Charlie's Pond is their long-term nature. We recently moved away from the frog parasites and have begun to examine another group of hosts in the pond. A few years ago, one of my graduate students, Giselle Broacha, looked at the population biology of two trematodes in the mosquitofish, *Gambusia affinis*. Mohamed Meguid, an Egyptian who did his Ph.D. with me, used the acanthocephalan, *Neoechinorhynchus cylindratus*, to do some interesting work on histopathology in green sunfish. More recently, Joel Fellis, another of my graduate students, examined the parasite fauna of centrarchids, primarily bluegill and green sunfishes, in Charlie's Pond and his work is now complete. Over a period of 9 months, he identified 16 species of helminths and counted some 37 000 worms! If we include the 10 species of helminths in the snails and frogs that do not cycle through fishes, we are now up to 26 species of helminth parasites in our small pond. This is quite an assemblage, especially considering we have not examined all of the vertebrates in the pond.

It will be of interest to see what direction our saga of Charlie's Pond will take over the next several years, and what else will flow in our "stream of consciousness."

References

Esch, G. W. and Fernandez, J. C. (1990). *A Functional Biology of Parasitism: Ecological and Evolutionary Implications*. London, UK: Chapman and Hall.

Ewing, S. A. (2001). *Wendell Krull: Trematodes and Naturalists*. Stillwater, Oklahoma: Oklahoma State University, College of Veterinary Medicine.

Rothschild, M. (1938). *Cercaria sinitzini* n. sp., a cystophorous cercaria from *Peringia uluae* (Pennant 1777). *Novitates Zoologicae*, XLI, 42–57.

Sih, A. (1984). The behavioural response between predators and prey. *American Naturalist*, **123**, 143–50.

Selected readings

Crews, A. and Esch, G. W. (1986). Seasonal dynamics of *Halipegus occidualis* (Trematoda: Hemiuridae) in *Helisoma anceps* and its impact on the fecundity of the snail host. *Journal of Parasitology*, **73**, 646–51.

Esch, G. W., Wetzel, E. J., Zelmer, D. A., and Schotthoeffer, A. M. (1997). Long-term changes in parasite population and community structures: a case history. *American Midland Naturalist*, **137**, 369–87.

Goater, T. M., Shostak, A. W., Williams, J. A., and Esch, G. W. (1989). A mark–recapture study of trematode parasitism in overwintered *Helisoma anceps* (Pulmonata) with special reference to *Halipegus occidualis*. *Journal of Parasitology*, **75**, 553–70.

Krull, W. A. (1935). Studies on the life history of *Halipegus occidualis* Stafford, 1905. *American Midland Naturalist*, **16**, 129–42.

Kuris, A. M., and Lafferty, K. D. (1994). Community structure: larval trematodes in snail hosts. *Annual Review of Ecology and Systematics*, **25**, 189–217.

Zelmer, D. A. and Esch, G. W. (1998). The infection mechanism of the cystophorous cercariae of *Halipegus occidualis*. *Invertebrate Biology*, **117**, 281–7.

4

Douglas Lake: early field parasitology in North America

An institution is the lengthened shadow of one man.

RALPH WALDO EMERSON, *ESSAYS: FIRST SERIES* (1841)

The great man is the man who does a thing for the first time.

ALEXANDER SMITH, *ON THE IMPORTANCE OF A MAN TO HIMSELF* (1830–67)

Field parasitology in North America, from the standpoint of both teaching and research, had its beginning at the University of Michigan Biological Station (UMBS) close by Douglas Lake, and near the tiny village of Pellston in the northern tip of lower Michigan. Research in field parasitology at UMBS began in 1914, with the work of William Walter (Will) Cort. The first of some 350 parasitology papers originating from UMBS was published by Cort in the *Journal of Parasitology* in 1915 (Cort, 1915a). He did not write all of the other 349 papers that followed, only about 20% of them! When one examines the full range of his publications from UMBS, there are two areas that represent major contributions to the development of modern field parasitology, and even epidemiology, in North America. The first was an effort to quantify the probabilities of success in trematode transmission biology. In addition to its ecological significance for the study of trematode community biology, it is symbolic because it represents the transition point of field parasitology research from what can be described as "natural history" to modern quantitative ecology. It was years ahead of its time in this regard. The second involved the identification of schistosome cercariae as the causative agents of swimmer's itch, still a vexing skin disease for humans in that part of North America. In 1926, George R. LaRue, the Director of UMBS, invited Will Cort and Lyle J. Thomas to teach a course in helminthology at the Station. Cort handled all the lecturing and Thomas concentrated on the laboratory. These two, plus LaRue (for inviting them), must be given joint credit for teaching the first field parasitology course at UMBS and thus in paving the way for field parasitology to be taught at biological stations in North America for the balance of the twentieth century. For these reasons, and several others that will become clear, Will Cort and UMBS are the focus of this essay.

Several years ago, I received a letter from Sterling Brackett, offering to give me some back issues of the *Journal of Parasitology*. He and his wife, Helen, were living in a retirement community near Chapel Hill, and graciously indicated their willingness to drive over to Winston-Salem and personally deliver his copies of the *Journal*. My students and I took them to lunch at Simo's, a local pub near Wake Forest. While eating our cholesterol-laden Lexington barbecue, we had a far-ranging discussion. It was there that I learned Helen Brackett was a daughter of Nellie and Will Cort. Later, I was to find that Mrs. Brackett's sister, Margaret (Peggy), was the wife of Louis (Louie) Olivier, who, like Brackett, was one of the outstanding parasitologists in the USA in the middle part of the twentieth century. Unfortunately, by the time I began writing the essays for this book, both of their husbands had died. However, the two lovely ladies agreed to meet with me and reminisce about their mother and father, their husbands, and about their experiences at UMBS. So, I arranged to spend some time with them at their retirement home down in Chatham County, south of Chapel Hill, North Carolina, in September of 2000. My good friend, Dick Seed, joined me for the second interview in March of 2001. I taped the interviews, each of which lasted about 2 hours. During the wide-ranging talks, we asked them all sorts of questions about people, places, and parasites having to do with their father and UMBS, and they shared themselves completely. They were charming beyond description.

When I began doing the research for this essay, I discovered that the first Europeans, probably French or English trappers/traders, traveled into the Douglas Lake country of Michigan in the early seventeenth century. At that time, Ottawa Indians occupied the land. It was a rich mixture of lakes, ponds, bogs, moraines, and streams, largely reflecting the impact of glaciation that had occurred there some 10000 years previously. The virgin forests in this region of Michigan at the time of the European incursions included exceptional stands of both deciduous and coniferous trees that, nearly 250 years later, were to provide marvelous opportunities for entrepreneurs interested in exploiting the resource for lumber. Between 1875 and 1885, one of these entrepreneurs, Mr. William

Pells, acquired approximately 27 000 acres of the forests in the vicinity of Douglas Lake. Eventually, Charles and Hannah Pells Bogardus inherited the property. Charles Bogardus had been a colonel in the Union Army during the Civil War and then a prominent politician and banker in Illinois. He and his wife moved to Michigan from Illinois in 1901, built a railroad, and started into the land and lumber business. Unlike his father-in-law, however, Bogardus apparently was not a very good businessman and became overextended, to the tune of some 12 million dollars! In 1908, the University of Michigan was looking at a site for a summer civil engineering camp near Douglas Lake and approached Charles and Hannah with the idea of leasing the property. The Bogardus countered by offering the University 1440 acres as a gift. In addition to engineering, they suggested it should be used for scientific purposes, and thus was born the University of Michigan Biological Station. Jacob Reighard, Professor of Zoology at the University, was named the first Director in 1909.

One of the prime players in the development of UMBS as a first-rate field station was George R. LaRue, who was born in Iowa in 1882. After finishing at Doane College in Blair, Nebraska, in 1907, he went on to do graduate work in parasitology at the University of Nebraska-Lincoln with Henry Baldwin Ward who was then head of the Zoology Department. When Ward left Nebraska to go to Illinois, LaRue went with him and received his Ph.D. there in 1911. He was then offered a position in the Zoology Department at the University of Michigan in 1912 with an annual salary of $1200. He accepted it, and ended up staying at the university until he retired in 1952. In June of 1912, he and his wife took a train from Ann Arbor to the tiny village of Pellston, named for William Pells, the original owner of the UMBS property near Douglas Lake. That summer, LaRue began a relationship with UMBS which was to last until 1939. He was appointed Director of the Station in 1916. LaRue was a fine parasitologist in his own right, but he was also a very effective, albeit highly irascible, administrator, and accomplished a great deal in shaping UMBS in its early days.

I always considered myself quite lucky because I was fortunate in having met Will Cort in 1963 when I was a National Institutes of Health post-doctoral trainee in the Department of Parasitology, School of Public Health, University of North Carolina (UNC)-Chapel Hill. Will, and his wife Nellie, had retired to Chapel Hill from the School of Hygiene and Public Health at Johns Hopkins. During my 2-year stay at UNC, Cort would come by the department on occasion, frequently stopping at my office where, among other things, he would regale me with tales of his youth in Colorado Springs, Colorado. I discovered he was a graduate of Colorado College (CC), class of '09. Since I too was a graduate of CC, class of '58, we had something in common from the beginning and we used to talk about the "olden" times, his and mine, when we lived in beautiful Colorado Springs. He told me of his teenage days, when he worked as a muleskinner, driving a wagon in nearby Cripple Creek, the site of the last, and the greatest, gold strike in Colorado. While at CC, he played end on the school's football team for 3 years, and on teams that twice defeated the University of Colorado – no small achievement for the tiny college at the foot of beautiful Pike's Peak. Most importantly for Cort though, he came under the influence of a biologist at the College, Professor E. C. Schneider. It was Schneider who encouraged him, then helped him obtain a graduate assistantship at the University of Illinois in 1910 where he was to work toward his Ph.D. under the direction of Henry Baldwin Ward. Cort had been tempted to major in English, but Schneider inspired Cort into following academic parasitology as a career.

After receiving his Master's degree with Ward, Cort returned to Colorado College in 1912–13 as a visiting professor and assistant football coach. Among other things, he needed to earn some money since he and Nell Gleason were to be married the next year. During that academic year, Schneider offered Cort a permanent teaching position at CC and he was sorely tempted to accept it. His temptation was exacerbated not only by his love of the Colorado Springs area and his genuine attraction to CC, but also by his affection for Professor Schneider. Another reason for leaving Illinois and going to CC was his apparent unhappiness with Ward. Nellie Cort related to her daughter, Peggy Olivier, "It was a good thing that he did go, because it gave him a rest from Dr. Ward. Dr. Ward was a fine teacher, but he was very, very, strenuous and arbitrary." She continued, "It was a great rest for Will to go back to Dr. Schneider who was just a sweet person and a splendid scientist too." After the year at CC, Cort returned to Illinois to prepare for his impending marriage to Nell. After the wedding, he planned to take his new bride back to Colorado Springs and resume his teaching career at CC. However, on the day before they were to be wed, Nellie's parents asked if they would like to house-sit their home in Champaign for a year, instead of returning to CC. When he heard the offer, Cort made an immediate decision to stay at Illinois and finish his Ph.D. He rushed to the administration building that afternoon and submitted his application for readmission to graduate school, with some 30 minutes to spare, according to Nell. He received a scholarship, the grand sum of $40/month, and that is what they lived on in their first year of marriage.

I did not know Ward, and I do not recall that I ever spoke with Cort about him. I came along so much later. However, I did talk with Cort's daughters about this parasitology icon of the early twentieth century. They concurred with Nell that their father felt Ward was "a difficult man to work for." Despite the stern mentoring, they still felt that their father believed Ward was an excellent person with whom to obtain his Ph.D. Whatever other feelings he may have had for Ward, it is also quite apparent that there was great mutual professional respect between the two men. For example, Ward started the *Journal of Parasitology* in 1914. For many of the next 16 years, he involved Cort with various aspects of the *Journal*'s operations, prior to Ward turning it over to the American Society of Parasitologists (ASP) in 1930. Indeed, Cort played an important role in the *Journal*'s eventual transfer to the ASP, and I have wondered if the transition would have gone as smoothly if Cort had not been involved.

I am positive Cort would say, and I know his daughters would agree, the best thing that happened to him while he was at Illinois was meeting Nellie Gleason at the University's short-lived Quiver Lake Biological Station in the summer of 1910. After he and Nell were married, and with his Ph.D. in hand, they headed for his first tenure-track teaching position at Macalester College, a small Presbyterian school in St. Paul, Minnesota. Before going to Minnesota, Henry (Allan) Gleason, Nell's brother and the Acting Director of UMBS, invited them to spend their summer in 1914 at the Biological Station on beautiful Douglas Lake. Gleason had just returned to the University of Michigan from a sabbatical that had allowed him to study botanical gardens around the world. According to Peggy Olivier, "he had kept a hand-written journal which ran to a huge length and he paid Mother $50 to type it all for him that summer. That paid for her board, which was the only out-of-pocket expense." She continued that Will "had some kind of small research grant which paid his way." So, it was in the summer of 1914 that Cort began his research on larval trematodes in Douglas Lake and environs, brilliant work that was to continue, almost without interruption, for more than 40 years!

After that first summer at UMBS, the Corts moved on to Macalester where he was a faculty member for 2 years. He and Nell enjoyed life at the small school, although he was to say later, "I'm afraid that my lectures on personal hygiene to freshman men and my course in Systematic Botany left much to be desired." They returned to UMBS in the summer of 1915. Then, after 2 years in Minnesota, they departed and headed for the University of California at Berkeley where he became Assistant Professor of Zoology. My impression from his autobiography and from discussions with his daughters was that the family was not too happy out there on the west coast. The cost of living was exceedingly high during World War I, and a heavy workload at the university undoubtedly contributed to their discontent. Moreover, his boss at Berkeley was the overbearing C. A. Kofoid, with whom he apparently did not get along very well. As pointed out by his daughters, while he was at the University of California, he began working on hookworm disease in, of all places, gold mines! When I first heard this, I wondered about the idea of doing hookworm research in a gold mine? However, the more I thought about it, where could you find a better site to do this kind of work? The deep mines are warm, they are moist, and there was a captive audience of miners who obviously could not go the surface every time they needed to relieve themselves. A hookworm's life cycle was perfectly matched for at least some of the gold mines of California.

Although Cort and his family did not return to Douglas Lake during their 2-year stay at Berkeley, the early UMBS connection was to pay the first of many long-term dividends. In the summer of 1915, he had met the great Robert Hegner at the Station and, in 1918, Hegner offered Cort a position at the Johns Hopkins School of Hygiene and Public Health, which he quickly accepted. Arriving at Hopkins in the fall of 1919, Cort took charge of teaching and research in helminthology, which, as he described it, was "the study of parasitic worms, not theology." He was to spend the rest of his professional career there, except for those wonderful summers at his beloved field station in Michigan.

At Hopkins, Cort quickly gained international recognition for his work on hookworm disease, traveling extensively to Puerto Rico, Panama, and China. However, it was almost like he lived a double life in terms of his research interests. His hookworm research was based at Hopkins, but larval trematodes at the field station in Michigan had a strong hold on him as well. Indeed, after talking with his daughters and examining his publication record, I feel the UMBS was a real "clarion call." The position at Hopkins was ideal for Cort and he enjoyed it tremendously. The summers at the UMBS though were, as he described, "very important in my career," a sentiment strongly shared by his daughters. He went on to say, "some of my best graduate students at Hopkins were first contacted at the Biological Station."

In 1923–24, he took his entire family to China, part of the travel I suppose he dreamed about when he proposed to Nellie back in Champaign in 1911. He, Nell, and their three daughters, aged seven, five, and three, traveled first by train from Baltimore to Colorado Springs and on to San Francisco, where they were joined by Norman Stoll, one of Cort's students at Hopkins. Then, it was by steamer to Yokahama and Shanghai, and finally by train to Peking, where Cort would teach at the Peking Union Medical College, and continue his research on hookworm disease. While in Peking, they lived in the home of Dr. and Mrs. Ernest Carroll Faust. Cort and Faust were on a faculty exchange, arranged by the people at Peking Union Medical College who, according to Peggy Olivier, "wanted to give Faust some additional experience." Fortunately, life in China was good and the two Cort daughters whom I interviewed had fond memories of the long and exciting trip so many years ago. They were there about 14 months before returning home.

In 1926, while working on hookworm disease in Panama, Cort received a message from LaRue, inviting him to teach a course in helminthology at UMBS. He jumped at the opportunity. Cort's only quid pro quo was that he have Lyle Thomas of the University of Illinois join him as his assistant – a relationship, or partnership if you will, that was to last for the next 16 years. Cort was to do all of the lecturing in the helminthology course and Thomas ran the labs. They began teaching their field parasitology course in the summer of 1927.

A brief aside is in order at this point. During his trips to study hookworm disease, Cort was usually accompanied by one or more other parasitologists of the time. The distinguished Maurice C. Hall, who was a researcher at the United States Department of Agriculture (USDA) in Beltsville, was a companion on at least one of these excursions. Hall, among other things, apparently loved to place these experiences in the context of sometimes witty, and sometimes quite serious, verse. While I was writing this essay, Dick Seed gave me a neatly typed volume, entitled *An Epic of the IHB*, written by Hall who had then given it to Cort. I presume it was intended as a gift because Cort's name was written on the cover. It was apparently left to the School of Public Health in Chapel Hill after Cort died, and came into Seed's possession when he was Head of the Department of Parasitology. In this little book, Hall described a long boat ride to Central America with several parasitologists, including Cort, and many of their experiences while there doing research on hookworm disease. Among the others who accompanied Cort and Hall on that trip was Norman Stoll, along with W. A. Riley, D. L. Augustine, W. H. Sweet, and D. M. Molloy. I checked with Peggy Olivier, but she did not know the meaning of IHB, except to suggest that it might have been the "International Health Board." The 100+ page volume written by Hall includes several brief poems in which he describes each of his colleagues. What follows is his description of Will Cort. Based on what I now know about Cort, I think it fits him well.

> The famous W.W., one-time Bill,
>   A speedy end upon the varsity,
> In Colorado's mountains got a start
>   As "one of Schneider's men" at old C.C.
> As a graduate he worked at Illinois
>   And chased the larval tremas to their lair;
> He took their excretory systems out
>   And placed a formula symbolic there.
> Then to Macalester and Berkeley's hills;
>   The Californian hookworm he pursues,
> And now Johns Hopkins puts him on her roll,
>   He fills the helminth-professorial shoes,
> And anything he fills, he fills it full,
>   For hc and Sweet make up our heavyweights,
> And either one would break the I.H.B.
>   If traveling was charged on poundage rates.
> When in his office, lab or lecture room
>   This Cort's a very serious one,
> But in the field he lapses now and then
>   And likes a little joke or tale or fun.
> Our expedition is the fourth he's led
>   From Baltimore to foreign lands afar,
> To Puerto Rico, China, Trinidad,
>   He goes wherever man and hookworms are;
> And where he goes, he brings the bacon home,
>   And printing presses groan beneath their load.
> A steady not too impulsive man,
>   He watches his step and knows his road.

The Corts bought their first automobile in 1929 and decided to drive it to UMBS for the summer of research and teaching. Think about it. In those days, there were no superhighways; indeed, outside cities and towns, there were few paved roads, most of them being dirt and gravel. Concrete highways were under construction all along their route, however, creating many long delays and frequent detours as they traveled. Gas stations were few, and motels were nonexistent. Planning a trip of this extent took great care, and making it required much patience, especially with three giggling young daughters in the back seat all the way from Baltimore to the northern part of Michigan! That summer of 1929 it took them five full days to make the journey.

They usually arrived at Douglas Lake about a week before the summer session started, and moved into their exceedingly tiny lodgings. During the first few summers, the family lived in a small, 12 × 12-foot cabin, and took their meals at a common mess hall. A few summers later, two of these small cabins were joined together, doubling the size of their accommodations for faculty (Fig. 25). Female and male students, faculty, and newly married couples were, respectively, segregated into Ladysville, Mansville, Facultyville, and Blissville. The two Cort sisters had a good laugh as they described the single water faucet in the street of Blissville (Fig. 26), where everyone had to go to brush their teeth on the sometimes very cold mornings in the north of Michigan. The cabins were heated with wood stoves. The family washing was done using a scrub-board in a metal wash tub,

Figure 25   Nell and Will Cort, at home (circa 1930s) in their University of Michigan Biological Station (UMBS) cabin. Will and his wife, Nell, began going to UMBS in the summer of 1914. Cort must be considered as one of the true pioneers in field parasitology in North America. (Courtesy of Mrs. Margaret Olivier.)

Figure 26   Blissville at University of Michigan Biological Station. These were the living quarters for newly married students. (Courtesy of Mrs. Margaret Olivier.)

then rinsed in the lake, and hung on clotheslines outside the tiny cabins. Water for washing clothes was heated in a bucket on the wood stove. When they wanted to bathe, they swam in the cold lake, as there were no showers in those early years. Mrs. Brackett and Mrs. Olivier both enjoyed the primitive living, and the two sisters exclaimed, "we had the best time!"

Sterling Brackett did his Ph.D. with C. A. Herrick at the University of Wisconsin. Herrick had been a student of Cort at Hopkins. Brackett became interested in UMBS from a high-school friend who studied conservation at the station one summer and recommended the place as a great experience (Fig. 27). It was there that he met Helen Cort, and they were married in 1935. Louis (Louie) Olivier was a student of Horace Stunkard at New York University and met Peggy Cort at the Station in the summer of 1935. Stunkard had taken his Ph.D. with Ward at Illinois and was a contemporary of Cort. By 1937, Louie proposed marriage and Peggy had accepted. Stunkard was a rather straight-laced man, very proper indeed. He strongly insisted that no one should get married until the husband-to-be had at least $5000 in the bank, which effectively ruled out graduate students (from this perspective, the financial status of graduate students has not changed much, e.g., always poor!). Peggy and Louie persisted, and her father even interceded on their behalf. Finally, Stunkard relented, and they were wed in 1938. Olivier worked as Cort's research assistant during

Figure 27   A limnology class working in Douglas Lake at University of Michigan Biological Station. (Courtesy of Mrs. Margaret Olivier.)

the next four summers. Peggy Olivier said that after she and Louie were married, they lived in one of the tiny, 12 × 24-foot cabins in Blissville, where all of the young couples were housed. Cabin size is irrelevant for a newly married couple at a field station, something to which my wife, Ann, and I can also attest. Early in our own marriage, during a hot summer in 1960 at the Oklahoma University Biological Station, we and our 9-month-old son, Craig, were assigned a 16 × 16-foot prefabricated World War II plywood shack, complete with no running water or air conditioning, lots of cockroaches, an occasional field mouse, and paper-thin walls. But this was OK, we were together and that's all that really mattered.

Cort wrote in his doctoral dissertation, which was subsequently published in the *Illinois Biological Monographs* in 1915, "Practically nothing is known of the life-histories of trematodes in North America" (Cort, 1915b). It can be comfortably said that knowledge of the basic biology of most parasites and parasite life cycles in those early days was exceptionally rudimentary. None the less, the concepts of a definitive and an intermediate host had been established, and it was understood by then that the transmission of some parasites was linked to predator–prey interactions, and that others were not. Gottlob Kuchenmeister had successfully determined the life cycle of the cestode, *Taenia solium,* in 1851. Joseph Leidy in the USA and Richard Owen in the UK had laid the groundwork for Rudolph Leuckart and Rudolph Virchow to complete their life-cycle studies on the nematode, *Trichinella spiralis*, in 1857. Algernon Phillips Withiel (APW) Thomas and Rudolph Leuckart had independently completed most of the life cycle of the trematode, *Fasciola hepatica*, in 1881–83. Ronald Ross received the second Nobel Prize in physiology and was knighted in 1902 for discovering the presence of *Plasmodium* parasites in anopheline mosquitoes in 1897. So, by the time Cort arrived at UMBS, life-cycle patterns in three of the major groups of parasitic helminths had been elucidated, and the mystery of malaria transmission was well on its way to being resolved. Even though some very important information regarding basic parasitology had been developed, the work that Cort, his students, and other colleagues accomplished at UMBS was highly significant. And whereas their empirical research was of enormous consequence, I think it is equally important that they were to establish, for the first time, the value of field parasitology in both teaching and research.

Cort did his doctoral research at Illinois on larval trematodes, and this work was to continue at UMBS. This region of North America is home to the great marl lakes and ponds characteristic of water high in calcium. Calcareous soils and waters are conducive to a rich molluscan fauna, which automatically translates into a diverse trematode fauna. Moreover, the Michigan landscape is cluttered with both lentic and lotic habitats due to glaciation that had occurred in the Great Lakes region. Minnesota claims to be the "land of a thousand lakes," but almost the same can be said for both Wisconsin and Michigan. The opportunities for research on trematodes and their biology were almost limitless at UMBS, and they were to be fully exploited by Cort and the others who followed over the next 85 years.

It is not my intention to describe each parasitology paper that came out of UMBS; after all, nearly 350 were published between 1915 and 2000, A number of authors, many of whom were students of Cort, made highly significant contributions through their research and these publications. Some of these outstanding parasitologists included Don Ameel, Marlowe and Florence Anderson, Paul Beaver, Harvey Blankespoor, Sterling Brackett, Eli Chernin, Dominic DeGuisti, Chester Hughes, Kathleen Hussey, Bruce Lang, Allen McIntosh, Don McMullen, Darwin Murrell, Louie Olivier, Lew Peters, Lyle Thomas, John Van Haitsma, and Anne Van der Woude, among others. As I looked through the many publications from UMBS, I was struck by their diversity. Don Ameel, Anne Van der Woude, and Cort, for example, published a long series of papers on the germinal development of several larval trematodes in their molluscan intermediate hosts. The Andersons detailed the biology and life cycles of several species of *Proterometra*. These are azygiid flukes that possess huge cercariae which resemble mosquito larvae in their size and behavior, making them inviting morsels for their piscine definitive hosts. Paul Beaver, who was later to gain international fame at Tulane for his work on creeping eruption and other nematode-producing diseases, published several papers on the morphology and life cycles of echinostomes and psilostomes. Harvey Blankespoor, presently a Professor of Biology at Hope College in Michigan, has had a long-standing interest in swimmer's itch in Michigan. To my knowledge, Harvey is the last in the lineage after Cort to teach field parasitology at UMBS. Eli Chernin, who was to establish a reputation for his later work on molluscs, conducted research at UMBS on the epizootiology of *Leucocytozoon simondi,* the potentially lethal protozoan, which is vectored to certain avian hosts via biting black flies. Chester Hughes contributed a number of papers on the biology of strigeid flukes before turning his attention to monogeneans during a long career at Oklahoma State University. Don McMullen was a student of Cort at Hopkins, but he published several papers on the life cycles and biology of plagiorchiid flukes in the Douglas Lake region. He was a contributor to the important trematode ecology paper in 1937, with Sterling Brackett and Cort, about which I will write in some detail later in this essay. McMullen was to become President of the ASP, but, tragically, died in office in 1967, to be replaced, ironically, by Louie Olivier, a close friend and Cort's son-in-law. Olivier was Vice President of the Society at the time of McMullen's untimely death. Lew Peters was a student of Ray Cable at Purdue, but also had spent time at UMBS. I first met Lew in 1960 when he came down to Oklahoma University Biological Station (OUBS) on Lake Texoma to help Professor Self, my mentor, who was attempting to work out the life cycle of the didymozoid fluke, *Nematobothrium texomensis*. They were not successful. While at UMBS, however, Lew was successful in completing the life cycle of *Allocreadium neotenicum*, an enigmatic progenetic fluke occurring as an adult in Douglas Lake water beetles. Of course, Lyle Thomas was Cort's partner in teaching helminthology at UMBS. Thomas' research focused on the biology and life cycles of both cestodes and trematodes, including *Halipegus eccentricus*, a hemiurid fluke that my students and I have studied over the years (see Chapter 3).

Cort published the very first parasite paper from UMBS in volume 1, number 4, of the *Journal of Parasitology* in 1915 (Cort, 1915a). In this brief note, he described the presence of *Gordius* larvae in the parenchymatous tissues of the trematode, *Brachycoelium hospitale*, in the gut of a green newt, *Diemictylus viridescens*. He suggested the infection was accidental, which it probably was, but he was probably also unaware that *Gordius* larvae frequently use snails as hosts and that the parasite was more than likely acquired by the trematode during its intramolluscan development, prior to transmission to the newt. Cort went on to write or co-write some 64 papers from UMBS during his 45-year association with the field station at Douglas Lake. I am familiar with much of his work, but there are two pieces of research in which he truly broke significant new ground. One was the identification of nonhuman schistosome cercariae as the cause of swimmer's itch. The second was the snail–trematode ecology paper he published in the *Journal of Parasitology* with Don McMullen and Sterling Brackett in 1937 (Cort *et al.*, 1937).

Many present-day biologists view natural history in an almost disdainful manner since it is not quantitative in scope. Personally, I feel that the natural history approach has an important place in modern biology and parasitology, and that it should not be disparaged. For a parasitologist, the ultimate in natural history research is the elucidation of a parasite's life cycle. If one goes back to issues of the *Journal of Parasitology* published in the early part of the twentieth century, the life cycles of many parasites are described. A number of authors stand out as consistent contributors in this arena, e.g., George LaRue, Horace Stunkard, Harley Van Cleave, Harold Manter, Asa Chandler, Wendell Krull, and many others. All of these men were also great naturalists. What is it that makes the great naturalist, that person with the special skill of being able to decipher nature's (sometimes cryptic) pattern? Miriam Rothschild, in an interview in *Scientific American* in 1996, is quoted as saying, "I think if you had to describe any talent I have, it is that I am a good observer. And that means that you don't only notice things, but you think about what you have noticed." All of these great parasitologists, including Will Cort, from the early part of the twentieth century had this same talent. Nowhere was this ability better shown by Cort's fieldwork than in his discovery of the cause of swimmer's itch in the summer of 1927. By this time, he was aware of schistosome life cycles. Then, he made the critical observation of a naturalist. More importantly though, he linked this observation to the schistosome's life cycle, a simple but simultaneously elegant union.

The life cycles of schistosomes include just two hosts, a molluscan first intermediate host and a mammalian/avian definitive host. Within the snail host, trematode development leads to the production of furcocercous, or fork-tailed, cercariae. On release from the snail, these cercariae swim vigorously for several hours. When contact is made with the surface of a definitive host, the cercaria attaches, then sheds its tail, penetrates the skin, enters the blood vascular system of its definitive host as a schistosomule, and, finally, makes its way to the host's liver where it stays for several days. On reaching this site, it must find a partner of the opposite sex with which it permanently pairs and, together, they migrate to an appropriate region of the host's venous system. Copulation and egg production follow. The exit process for schistosome eggs from blood vessels is quite complicated and outside the remit of this essay. However, the manner in which this parasite can first stimulate a host's immune response, then exacerbate and use it as a means by which eggs can escape from the definitive host, is hugely fascinating, and a wonderful object lesson in understanding how parasites can manipulate their hosts' immune systems.

Human schistosomiasis is a serious problem in certain regions of the world. There are three species that routinely infect humans in endemic areas, and probably a couple of others as well. The most recent estimates indicate that in excess of 200 million people are infected with the highly pathogenic schistosomes, with approximately 20 million showing morbidity and 20000 dying annually from the disease.

In view of Will Cort's tremendous interest in both schistosomiasis and hookworm disease, I was fascinated to read Peter Hotez's account of the impact of these two helminth parasites in the People's Republic of China (PRC). Writing in *The China Quarterly* (2002), Peter said that, "More than any other nation during the 20th century, human helminth infections have exerted their greatest impact on the People's Republic of China," and schistosomiasis has certainly been one of the more difficult problems in the PRC. I can personally recall one lecture in particular on this particular disease when I took introductory parasitology from Doc Stabler back in 1958. I was so impressed that I incorporated portions of it into my own lectures on the schistosomes when I first began teaching parasitology to undergraduate students in 1965. Stabler had alluded to the impact of *Schistosoma japonicum* on a significant world event of the time, one that may even have kept the USA from a military conflict with the PRC. In a letter to me regarding this and another essay that follows, Peter talked about the same event in even greater detail and it is worth repeating here.

In 1950, during Dwight Eisenhower's first term as President, the PRC almost certainly made the decision to invade Taiwan (= Formosa). Toward this end, several divisions of Mao's People's Liberation Army were moved into the provinces of Fujian and Zhejiang, opposite Taiwan, apparently in preparation for an amphibious assault on the island. But, then, mysteriously to the western nations at least, nothing happened. We have since learned that, ironically, Mao's divisions were themselves amphibiously assaulted by (you guessed it) cercariae of *S. japonicum*, to the extent that very large numbers of them were incapacitated by the parasite. In his book, *Mao's Way* (University of California Press, Berkeley), E. E. Rice (1972) referred to this parasite as "the fluke that saved Formosa." Peter Hotez wrote in *The China Quarterly* that, in the ensuing decade of the 1950s, "Mao mobilized hundreds of thousands of peasants to bury *Oncomelania hupensis* snails along tributaries of the Yangtze River. In some cases the peasants removed the snails one-by-one with sticks. Schistosomiasis control was a major component of his patriotic campaigns during The Great Leap Forward in 1958, and was even the subject of a famous poem he composed with the title, *Farewell to the God of the Plague.*"

Fortunately for most of the world, schistosomes of the sort that affected Mao's divisions are not a problem. There is, however, another kind of schistosomiasis that does have a worldwide distribution, but its pathogenicity for the human host is, shall we say, just "skin deep." When a schistosome cercaria enters the skin of its appropriate definitive host, there may, or may not, be a host reaction to the invading parasite. If there is, it is limited to the point of entry of the parasite. It consists of a localized hypersensitivity reaction that raises a small red papule which, simply put, itches! The papule may develop into a pustule (Fig. 28), which lasts for a few days, then disappears. There is some argument as to whether the skin reaction will occur when the cercaria penetrates the correct definitive host. In North America and other parts of the world where human schistosomiasis is not endemic, this question is moot. If a nonhuman mammalian or avian schistosome cercaria penetrates the skin of a human, then there will almost always be a reaction (there are a few exceptions), producing what is commonly called swimmer's itch. These nonhuman schistosome cercariae, when they penetrate the wrong host, are overcome by the host's immune response, killed in the process, and then resorbed. In North America, this problem is especially common in the north central states of the USA, in several of the provinces of Canada immediately to the north of these states, and along the east coast, where it is sometimes called "clam-digger's itch."

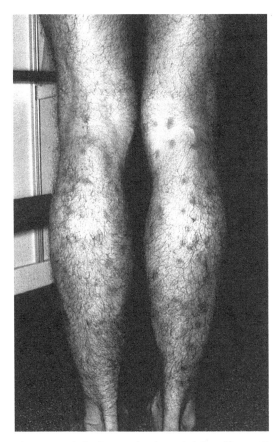

Figure 28 A classic case of swimmer's itch. Will Cort discovered that avian schistosome cercariae were the cause of this (still vexing) disease in humans. (Courtesy of Russ Hobbs.)

Why the fuss if the only problem with infection by a schistosome cercaria is limited to an itch? If you are a resort owner on a lake, you want the local tourists to come and spend a few days swimming at your beach, or wading in shallow water on a hot summer day. The last thing you want is for your guests to complain about the awful rashes (swimmer's itch) they acquire while swimming or wading in your lake! Make no mistake about it either, if you own a resort on a lake with swimmer's itch, the word will get out, and the tourists will stop coming. Swimmer's itch can be costly for these resort owners because the tourists will definitely find another place to spend their leisure time, i.e., on a lake without swimmer's itch.

The discovery of schistosome cercariae as the causative agent for swimmer's itch is an interesting story, one that also strongly reflects the occasional case of serendipity in science. Moreover, it is a classic example of how one of the first epidemiological studies was accomplished in North America. Cort went to Douglas Lake each summer during 1914, 1915, 1916, 1920, and 1925. Every time, he developed a peculiar dermatitis on his ankles or wrists, particularly after collecting snails at a place called Sedge Pool, a small, permanent beach pool on the north side of Douglas Lake. In fact, the dermatitis he and many others acquired was first known as "Sedge Pool Itch." In 1916, virtually an entire class of invertebrate biology students developed the same kind of dermatitis after working for a morning at the Sedge Pool location. In the summer of 1924, two girls, wearing bathing suits, spent an entire day collecting at the Sedge Pool site. Both developed extremely debilitating cases of dermatitis, which in fact left them prostrate for several days, with swollen legs covered by large numbers of the pustules. Examples of a similar sort of dermatitis had been reported by workers collecting in other aquatic habitats in Douglas Lake and by resort owners on other lakes in the vicinity over the years.

The etiology of this dermatitis remained unknown until 16 July 1927. On that day, Cort went on a field trip to collect snails along the north shore of Douglas Lake (the following account is adapted from an article written by Cort in 1928 that appeared in the *Journal of the American Medical Association*). As was the usual procedure, snails collected from the lake's substratum were tossed into a bucket containing several inches of water. When he returned to the lab, he isolated the snails into half-pint milk bottles. In doing the sorting, his left hand was constantly in contact with water in the bucket. After about 10 minutes, the wrist on this hand began to develop a prickling sensation, something like mosquitoes would cause as they bite. He quickly washed the wrist in alcohol and the sensation of biting insects subsided. However, about 1 a.m. the next morning, he was awakened with an intense itching on his left wrist. Later that morning, prominent red papules had developed on his wrist and the itching persisted, becoming almost unbearable at times. Between 48 and 72 hours after sorting through the snails, the papules became pustular, his hand and wrist were both badly swollen, and the itching was excruciating. After 4 days, the swelling had subsided, and the itching was much less intense. The pustules then began to dry up and, after a couple of more days, disappeared completely.

Cort had just experienced a classic case of swimmer's itch. He had had the same dermatitis before, but this time the itch was associated directly with water from which he had been removing the snails for isolation. Since he knew that skin-penetrating schistosome cercariae were present in snails of the region, he immediately reasoned that cercariae might be causing the dermatitis. When he checked his field notes, he discovered that the only parasites being shed from these snails were *Cercaria elvae*, a common schistosome in the area. So, he purposely placed some more of these cercariae on the back of his wrist. As the water dried, he began to experience the same prickling sensation that he had had on 16 July, and previously as I have indicated. He developed the same sequence of signs and symptoms he had just gone through, this time though without any swelling. As the summer progressed, he conducted a series of experiments in which, for example, he demonstrated that a number of snail species were involved with producing cercariae that caused swimmer's itch. He also found that humans varied considerably in their reaction when exposed to these cercariae. Some developed the whole range of signs and symptoms he himself experienced, whereas others were completely, or partially, refractory. I was amused when Cort's daughters told of their father coming home one evening and placing small drops of water containing cercariae on their arms to see if they were susceptible to the parasites. They were. He also determined that not all species of schistosomes have cercariae that can cause swimmer's itch.

When Cort went to Douglas Lake in 1914 the first time, he unknowingly encountered the cercariae that caused swimmer's itch (Fig. 28). In fact, he himself contracted swimmer's itch each summer he went to the lake, but he had no reason to suspect his "own" trematode parasites were causing the dermatitis prior to his discovery of the causative agent(s) in 1927. *S. douthitti* is now known as one of the most common of the swimmer's itch-producing cercariae in North America. Ironically, this parasite had been described and named by Cort in 1914 in his doctoral dissertation, before he ever set foot in Douglas Lake. In his dissertation, Cort wrote, "No suggestion can be made as to the life-history of *Cercaria douthitti*." He continued, "In fact hardly a suggestion has been made in regard to the life-histories of the forked-tailed cercariae and no experiments that I can find have been carried out to trace their development." Unknown to Cort, he was on the trail of discovering the cause of swimmer's itch when he described *C. douthitti*, being shed from *Lymnaea reflexa*, collected in a small pond from the suburbs of Chicago, Illinois. However,

Figure 29 A sign warning would-be bathers to stay out of the water because of the potential for acquiring swimmer's itch. (Courtesy of Al Bush.)

the solution of the swimmer's itch problem was to elude him for 13 more years.

As I mentioned earlier, swimmer's itch was, and still is, a nuisance for resort owners in the upper mid-western part of the USA, and in contiguous areas of Canada as well. There are three ways of dealing with the problem. The first is to avoid swimming in places where swimmer's itch is known to occur. Signs (Fig. 29) are frequently erected near ponds or lakes where swimmer's itch occurs, but on a hot day, a cooling swim is difficult to resist, no matter what a sign might say. The second alternative is to kill the infected snails. In the early days in Michigan, the molluscicide of choice was copper sulfate, and it is still used in some places. Over the years, many tons of this chemical have been dumped into a great many ponds and lakes in Michigan, Wisconsin, and Minnesota. It is effective, but the trouble with copper sulfate in high concentrations is that it kills many other organisms, including fish. For a resort owner in Michigan, the stench of dead fish on a swimming beach is just as bad as having swimmer's itch. Moreover, I heard a very well respected malacologist, Elmer Berry, say many years ago that trying to kill all of the snails in a lake was like trying to kill all of the poison ivy in the woods – it is senseless even to try.

The third course of action is to kill the definitive hosts, but this is a very dicey proposition, especially in these days of such strong antivivisectionism. More simply put, there are too many infected birds, and bird lovers. Recently, however, Harvey Blankespoor at Hope College in Michigan and several of his colleagues have developed a rather innovative procedure for the control of swimmer's itch. They have used a broad-spectrum anthelminthic drug, praziquantel, aimed directly at the adult parasites in their definitive hosts. In recreational lakes in that part of North America, mallard ducks and common mergansers are the most frequent definitive hosts for the schistosomes that cause swimmer's itch. In the experimental scheme developed by Blankespoor and his colleagues, a trapping program was initiated for both mallards and mergansers in a lake known for its swimmer's itch. They checked their captured birds for schistosome eggs being shed in the feces, then they fed the birds praziquantel in appropriate doses. The treated birds were released and as many as possible were recaptured at various intervals of time afterwards. Their results clearly show the efficacy of the drug for the avian schistosomes and suggest a potentially effective, though labor-intensive, method for the mitigation, if not control, of the swimmer's itch problem. I suspect Will Cort would be pleased.

Cort, McMullen, and Brackett were the first to employ a quantitative approach in field parasitology, making their research truly cutting edge at the time. In the broadest sense, their studies were designed to examine the prevalence of various species of trematodes in the snail, *Stagnicola emarginata anulata,* in Douglas Lake and nearby Burt Lake. They also made a concerted effort to ascertain the differences in the prevalence of infected snails, both temporally and spatially, at several different sites in these two lakes. The most innovative feature of their work, however, had to do with the manner in which they quantified their data. They determined the number, and nature, of multiple infections in individual snails, and then attempted to establish mathematically if certain combinations of trematodes in multiple infections had occurred more, or less, often by chance alone. In modern parlance, they had become involved in the first study of trematode infracommunities in snails, with the aim of identifying, or at least suggesting, the nature of the structuring forces in these infracommunities.

Douglas Lake is of moderate size, compared with others in Michigan. It has a surface area of approximately 15 km², a maximum depth of about 28 m, and a maximum length of 6.1 km. Throughout much of the early twentieth century, the lake had undergone eutrophication, but not to the point the hypolimnion became completely anoxic during periods of thermal stratification. The Cort *et al.* study of 1937 was focused in three localities of Douglas Lake and one in nearby Burt Lake. At all four sites, there were extensive littoral zones, capable of being easily waded. Each site also had scattered stands of emergent vegetation. After collections were made, snails were returned to the lab and isolated in half-pint milk bottles, which were then examined for cercariae using a hand lens. All snails were necropsied to confirm the presence or absence of infections. During the summers of 1935 and 1936, 7259 *S. emarginata angulata* were collected from the four study sites on 19 occasions and then examined for the presence of infections. A total of 21 species of trematodes had been identified in this species of snail since Cort began working in Douglas Lake in 1914. In the 1935–36 study, they found 17 of the 21 species. Prevalences of the trematodes ranged from 0.1% to a high of 50.9%. In all, there were seven strigeids, two schistosomes, six plagiorchiids, one echinostome, and one monostome. By far the most common cercariae shed were those of the strigeid, *Diplostomum flexicaudum*, a schistosome identified as *Cercaria stagnicolae*, and the plagiorchiid, *Plagiorchis muris*. *Diplostomum flexicaudum* uses gulls as definitive hosts and fishes as the second intermediate hosts. The life cycle of the most common schistosome, *C. stagnicolae*, is still not known, but they assumed it infected an avian definitive host, probably a gull. *P. muris* occurs in a range of avian and mammalian definitive hosts.

With such a large collection of snail hosts and the relatively high number of infections, there were fairly large numbers of multiple infections. The authors wanted to know whether these infections occurred by chance, or if there was any indication of negative interactions that might reduce the chances for mixed infections. They knew that the intramolluscan development of all species of trematodes in their study occurred strictly in the hepatopancreas, certainly creating an opportunity for trematode-trematode interaction. Although they did not know the life cycles of all their parasites, they knew that only two of the 17 species in their study used a redia in their intramolluscan development. These two parasites were *Echinostoma revolutum* and *Notocotylus urbanensis*. They observed that neither of these species co-occurred with any other parasite and, although they did not suggest that antagonism had occurred to create these single-species infracommunities, we now know that there is none the less a clear possibility for predation/competition by rediae of these two species. Several investigators, e.g., Armand Kuris, Wayne Sousa, and Kevin Lafferty (see also Chapter 3), have provided strong evidence over the years that this kind of negative interaction may result in dominance by trematode species with rediae in their life cycles.

The most common double infections included five combinations of parasites, none of which had a redia in its life cycle, i.e., strigeid/strigeid, strigeid/schistosome, strigeid/plagiorchiid, schistosome/plagiorchiid, and plagiorchiid/plagiorchiid. Triple infections involved nine different combinations. The lone quadruple included three strigeids and a schistosome. The group estimated that the number of double infections that would be expected through chance by multiplying the observed prevalences of each species of parasite times the number of snails in a given collection. For example, if the prevalence of *D. flexicaudum* cercariae was 40.9%, *C. stagnicolae* was 20.5%, and the total number of snails present in a given sample was 171, then the number of double infections involving these two species of trematodes by chance should be 14.

Three interesting generalizations emerged from their data analysis. The first involved the combination of *D. flexicaudum* and *C. communis*, which occurred 61 times, with the prediction that they should have occurred together 75 times by chance alone. Both of these parasites have similar life cycles, using gulls as their definitive hosts and fishes as their intermediate hosts. The similarity in life-cycle patterns of these two species, they asserted, would account for the parity in prevalence that was predicted and what they actually observed. The second observation involved *D. flexicaudum* and *C. stagnicolae,* a strigeid and a schistosome, respectively. In this case, the two parasites were seen in combination at all sites in Burt and Douglas Lake at very near the predicted numbers. In Burt Lake, however, in two collections, the double infections observed were far below those predicted by chance alone. The authors accounted for these deviations by suggesting that there had been differential mortality in snails harboring the two species. The third observation extended to several species in which there were consistent reductions in the number of combinations actually observed as compared to the number that was predicted. In their paper, Cort *et al.* indicated that "there must be an immunity produced by the penetration of the miracidium of one of these species which prevents infection of the same snail by the other; or that there is antagonism present which prevents development of the other when one is present." They also noted that "the immune or antagonistic relationships appear to be between certain individual species rather than between whole groups." They concluded, "there is no evidence from the present work that sheds any light on the mechanism of these immune or antagonistic relationships."

All of these findings were to have considerable impact on the way in which workers in later years have pursued questions regarding the diversity and the nature of structuring forces in intramolluscan trematode infracommunities. As near as I can tell, these studies were also among the very first to invoke the possibility of molluscan immunity, or of antagonism, in structuring trematode infracommunities in molluscan hosts. Although I have a number of years of experience working with the biology of snail–trematode systems, I felt that I should confirm this assertion. So, I called Timothy Yoshino, one of the premier investigators in the world working on snail–trematode interactions from an immunological perspective. Tim was familiar with the Cort *et al.* paper. When I read him the passages that I quoted above, he was in complete agreement that the Cort *et al.* paper was indeed "a classic" and that appropriate credit was due to these three researchers with respect to this particular contribution.

As mentioned above, another objective of the Cort, McMullen, and Brackett paper was to examine changes in the prevalence of parasites between 1935 and 1936. I think their data are somewhat disappointing in this regard. They simply did not have large enough sample sizes to justify any real conclusions, except to say there was variability in some places and not others. Most of the variability was exhibited by the rare species – precisely what might be expected. An interesting observation occurred in 1936, however, probably the consequence of one of the hottest summers ever recorded in the USA, let alone in Michigan, e.g., 5 days at UMBS in excess of 100°F over a stretch of 11 days in which temperatures were abnormally high! The elevated air temperatures apparently greatly impacted water temperature, especially in the shallow littoral zones of Douglas Lake. These high water temperatures were, they suggested, responsible for greatly accelerating the development of trematode larvae inside the shells of trematode eggs, as well as the development of trematode larvae in snails, resulting in distinctively different patterns of cercariae emergence between the two summers.

A few years ago, in 1997, as Editor of the *Journal of Parasitology*, I received a manuscript from Brian Keas and Harvey Blankespoor, two old friends of mine. In this paper, they reported on the results of their efforts to repeat the Cort *et al.* survey from 1937. Cort had also looked at one of these sites in 1960 to see what sort of changes had occurred since the 1937 study. Whereas the results of Keas and Blankespoor were inconclusive with respect to use of larval trematodes as indicators of habitat change, they showed that the numbers of species present had dropped from 21 to eight, continuing a trend observed by Cort in 1960. Moreover, the prevalence of the most abundant trematode had dropped from 50.9% to 13%. Most of the declining species were those that use gulls as definitive hosts. The only species exhibiting an increase in prevalence were those that use ducks as their definitive hosts. Keas and Blankespoor concluded that the reduction in parasite community diversity was the result of increased human habitation in the region, an explanation with which I strongly concur.

As I noted at the beginning of this essay, research in field parasitology in North America began in 1914 with Will Cort. It continues in North America, Europe, and elsewhere, at both freshwater and marine biological stations, and at various other field sites. Much of this work is quantitative in scope, but a great deal of it is still "natural history" in character as well. As I argued earlier, both methodologies have been, and still are, highly productive. In recent years, parasitologists, and others, have begun to use various molecular and genetic techniques in the study of parasites and host–parasite relationships. This is a good approach and must be fostered. Here, I would provide an admonishment, however. I have genuine fear that we will lose sight of the organisms involved, and become too focused just on the techniques. Research in parasitology has a tendency to be "trendy." Field parasitology, whether in the context of natural history, quantitative ecology, or molecular ecology/epidemiology, has much to contribute to our understanding of host–parasite relationships. If those working in this area can sustain a balanced approach, then, in my opinion, the future of field parasitology research is as bright now as it was when Cort first visited Douglas Lake almost 90 years ago.

In contrast, I am less than sanguine with respect to the present status of teaching field parasitology at biological stations. Cort and Thomas began it nearly 75 years ago at UMBS. I know for a fact that field parasitology was part of the summer curriculum for at least a dozen biological stations when I began doing this kind of research 30+ years ago. To my personal knowledge, however, it is now being taught at only a couple of biological stations, and I find this most unfortunate. There are several reasons for this teaching "meltdown." I remember at the W. K. Kellogg Biological Station (KBS) when one of the resident ecologists (who shall remain nameless) challenged George Lauff, who was then the Station Director, regarding his decision to include field parasitology in the summer curriculum at KBS. This person claimed that the course belonged on campus in the vet school, but George kept it going as long as he remained Director. Before coming to KBS as its Director, George had spent time at UMBS where field parasitology was still being taught and I suspect he saw its value there. When he retired several years ago, the course in field parasitology disappeared. I don't know why it was deleted from the curriculum, but I can assume. I guess what I am saying is that many present-day ecologists are still somewhat (sometimes even completely) ignorant (or perhaps confused?) with respect to the importance of parasites in the natural scheme of things. Fortunately, there are signs that this may be changing since, for example, parasites and parasitism are being given more than a passing nod in several of the newer ecology textbooks, but this is the only positive note I see at the present time.

One of the places where field parasitology is still being taught, with great enthusiasm and on a regular basis, is at the Cedar Point Biological Station, situated out on the high plains of western Nebraska. It is my intention to focus the next essay (Chapter 5) on this station and the way field parasitology is being taught there. The persons responsible are two well-known parasitologists, both old friends, John Janovy, Jr. and Brent B. Nickol. I had a chance to visit Cedar Point and sit in on one of John's classes in August of 2001, and I'm glad that I did. Not only did it give me some ideas for teaching my parasitology course here at Wake Forest, I was also refreshed by his approach, his enthusiasm, and his knowledge of the field in general. I also want to use the next chapter as a further tribute to both Will Cort and Lyle Thomas. After all, it was they who established the precedent and set the standard for all of us who have ever had the fun of teaching, or taking, a course in field parasitology at a biological station.

References

Cort, W. W. (1915a). *Gordius* larval parasites in a trematode. *Journal of Parasitology*, **1**, 198–199.

Cort, W. W. (1915b). Some North American larval trematodes. *Illinois Biological Monographs*, **1**, 1–86.

Cort, W. W. (1928). Schistosome dermatitis in the United States. *Journal of the American Medical Association*, **90**, 1027–29.

Cort, W. W., McMullen, D. B., and Brackett, S. (1937). Ecological studies on the cercariae in *Stagnicola emarginata angulata* (Sowerby) in the Douglas Lake region, Michigan. *Journal of Parasitology*, **23**, 504–32.

Cort, W. W., Hussey, K. L., and Ameel, D. J. (1960). Seasonal fluctuations in larval trematode infections in *Stagnicola emarginata angulata* from Phragmites flats on the Douglas Lake region, Michigan *Proceeding of the Helminthological Society of Washington*, **27**, 11–13.

Holloway, M. (1996). Profile: Miriam Rothschild. *Scientific American*, May, 36–8.

Hotez, P. J. (2002). China's hookworms. *China Quarterly*, **172**, 1029–41.

Keas, B. E. and Blankespoor, H. D. (1997). The prevalence of cercariae from *Stagnicola emarginata* (Lymnaeidae) over 50 years in northern Michigan. *Journal of Parasitology*, **83**, 536–40.

Rice, E. E. (1972). *Mao's Way*. Berekeley, California: University of California Press.

5

___

A day in the life of a field parasitology student, "Janovy style"

From the mountains, to the prairies . . .

IRVING BERLIN, *GOD BLESS AMERICA* (1938)

Nature sings her exquisite song to the artist alone, her son and her master – her son in that he loves her, her master in that he knows her.

JAMES ABBOTT MCNEILL WHISTLER, *TEN O CLOCK* (1888)

I do not know how Will Cort and Lyle Thomas taught their helminthology/parasitology course at University of Michigan Biological Station (UMBS). However, I suspect it was not unlike the one that John Janovy, Jr. and I took from Professor Self at the Oklahoma University Biological Station (OUBS). I used the same format when I taught field parasitology at the W. K. Kellogg Biological Station (KBS) each summer from 1965 through 1974. The lectures were a basic summary of general parasitology, but included some emphasis on several parasites and diseases of medical importance as well. After all, how can anyone teach parasitology, even a field course in North America, without dealing with aspects of malaria, trypanosomiasis, leishmaniasis, and other important protozoan parasites?

The laboratory part of the field parasitology course, however, is what set it apart from the general parasitology I have taught here at Wake Forest for the past 37 years. There were several objectives I had in mind at KBS. Among other things, I wanted to be sure my students had a strong sense that parasites are ubiquitous in nature and that their hosts are inclusive of almost every vertebrate and invertebrate animal they might encounter. I showed them how, and where, to collect and isolate snails, in part to give them an idea of the enormous diversity of, and reproductive potential by, the intramolluscan stages of digenetic trematodes. I attempted to give my students a feel for the way in which vertebrate animals should be collected, i.e., sanely and ethically. Students were taught the proper way to necropsy a vertebrate animal, beginning with the great care required in isolating the visceral mass once the animal is opened. The importance of carefully separating the organs, of making blood smears (after all, not all parasites are worms – many are protozoans!), and then of examining each organ separately, using a dissecting microscope, was impressed on them. In doing the necropsy, students learned that parasites in a vertebrate animal may be present in virtually any site, ranging from the brain, to the kidney, to the subcutaneous tissues, and even the vitreous humor or lens of an eye. Next, I wanted the students to understand that the biology of intermediate and definitive hosts is inextricably linked, in many cases by food chains and predator–prey relationships.

Once isolated, the parasite must be fixed, preserved, and stained; staining is almost an art form for some species of helminths. Then, there is the parasite's identification – not always an easy task for the beginning student. Many students are unaware of the binomial system of nomenclature, even though they may be senior undergraduates, or first-year graduate students. The reason for this gap in their knowledge is usually simple, but troubling. In many colleges and universities in North America, biochemistry or molecular biology has gradually replaced fundamental biology, which is a really great tragedy for beginning students. The result is that some very basic information must frequently be passed along in advanced courses. It is almost like teaching "remedial" biology to these students, a necessary, but wasteful, use of valuable time. For example, I recently taught several postgraduates in a nonmajor Principles of Biology course that is used by these students as part of a general program to enhance their chances for admission to medical school. One of these students was a graduate with a degree in molecular biology, and a B+-grade average, from a prestigious north-eastern university. I was astounded to learn that she had never been "inside" the carcass of an animal. She said that they started with DNA in the introductory course, and finished with it in the last course she took before graduating with her B.S. degree. Not that DNA or other aspects of molecular biology are unimportant, but this is not a good way to teach introductory biology!

On an even more personal note, we recently learned that our granddaughter, Morgan McCall, had been selected to attend the prestigious Governor's School in Richmond, Virginia. Morgan and I were on the telephone a few days ago, talking about what courses she would be taking in her first year. She was naturally excited. One of them she identified was Molecular Biology! Why in the world would they offer such a course in the first year? She did not have an answer and I don't either. My strong opinion is that introductory biology should be initiated with principles, i.e., basic chemistry, basic biochemistry, cell structure and function, mitosis and meiosis, Mendelian genetics, enough molecular genetics to understand how information is transcribed and translated, some organismic physiology, ecology, and evolution, including the Hardy–Weinberg law. It might be argued that this is too much. It isn't, especially if it is taught with rigor and enthusiasm. I have omitted any sort of plant or animal survey, but I contend that, with a basic understanding of these essential principles, a student should be able to handle survey courses or any other kind of advanced biology course that a modern department of biology could or would offer. I am not opposed to teaching molecular biology, far from it. But I strongly believe we would be turning out much better students, both biology majors (and probably more of them) and nonmajors, if they are first well grounded in the principles of general biology. In closing, this must sound like I am preaching. I guess that is what I am doing, so, Amen!

Back to field parasitology. Once a basic understanding of systematics/ taxonomy has been achieved, and the parasite has been appropriately stained and mounted on a slide, which diagnostic key should be used? How does one find and employ the primary literature? Another expedience for students at a field station is that they are usually isolated from their usual urban environment. Accordingly, both vertebrate and invertebrate animals, of all kinds, are available for necropsy. The method used in collecting might be a basking trap for turtles, a pellet gun for a robin, the familiar gill net for fishes in the nearby pond or lake, or driving the local roads at night looking for died-on-the-road (DOR) carcasses.

A great advantage of a field station is that, as a teacher, one has the students' complete attention. There are usually no TV sets at a field station, or English essays that must be written, or sorority/fraternity meetings or football games to attend. At a field station, there is only biology, giving the students the chance to become completely absorbed in their course work, with no distractions. The work is intense. There is, however, almost always one night each week during which the students leave the station and head for the local pub or tavern. This is where the students gather to discuss an hour exam taken earlier in the week and a "devilishly devious" essay question on the exam. Or, they might talk about the peculiar parasite removed from the urinary bladder of the bullfrog captured at a nearby pond, or rehash (usually by embellishing) a story about the professor (this is how legends are made). It is interesting that every field station has a local "watering hole," where students gather to eat some usually greasy food, down a few "brews," and interact socially. For us at Gull Lake, it was Gilkey's Tavern. For those at Cedar Point, it is either the Sip 'n' Sizzle, or Chere's. At Douglas Lake, it was Hoppy's. Actually, these places are sometimes more useful than just as watering holes. Darwin Murrell, an old friend and past President of the American Society of Parasitologists, told me that he met a fellow student, Joyce, at Hoppy's near UMBS at Douglas Lake. She was later to become his wife!

Sunday at a field station is the quiet day of the week. This day is for the family. It is used for lounging at the beach, or for catching a few rays on the floating platform near the field station's almost-always-present boat dock. At Gull Lake on Sunday afternoons, we were usually treated to a yacht race, with white-sheeted craft gliding swiftly over the sparkling blue-green water and young children wading or frolicking on the generally crowded beach. Sunday evenings would find many of the professors back in the lab writing their Monday-morning lectures or preparing materials for the next week's labs.

I have attempted to introduce this essay with a few ramblings regarding the field station experience in general terms, but what about the Cedar Point Biological Station (CPBS) in particular? To get to Cedar Point, I am first going to take you to the beautiful Rocky Mountains of Colorado. These mountains rise abruptly out of the high plains in western North America. Most of them along the Front Range just to the west of Colorado Springs are not high – around 10 000 feet. The exception is Pike's Peak (Fig. 30), a real anomaly because it rises to a height of 14 000+ foot. It was on top of this wonderful mountain, in 1893, that Katherine Lee Bates wrote the words for *America the Beautiful*. Pike's Peak was her "purple mountain majesty." She was immensely impressed as she gazed down on to the beautiful "fruited plains" of eastern Colorado – a great exaggeration (they are not "fruited"), but that's OK. If you have not driven the Pike's Peak Highway or ridden the cog railroad to the summit, you must do it because the view, in all directions, is breathtaking. It is easy to see why Bates was so impressed when she penned those elegant words. In *God Bless America,* written by Irving Berlin in the early twentieth century, there is a phrase, "From the mountains, to the prairies. . . ." I chose to use these words as an epigram for the present chapter, because the Platte River has its origin in the mountains of Colorado and then moves out on to the prairies of both eastern Wyoming and Colorado.

Figure 30   The beautiful Pike's Peak just west of Colorado Springs, Colorado, not far from the headwaters of the South Platte River. Pike headed a party that attempted to climb the mountain in the early nineteenth century. They were unsuccessful and claimed it would never be climbed. There is now a cog railroad to the top and a road as well. The latter is used for stock car racing to the top of the Peak each Fourth of July!

In no small part, this essay is about the Platte. This river is crucial to the existence of CPBS, to the field parasitology that is taught there, to the research done there by John Janovy, Jr. and Brent Nickol, and to all of the students who passed this way over the years. The Biological Station of the University of Nebraska is situated adjacent to Keystone Lake on the North Platte River, about 10 miles from the Keith County seat of Ogallala, on the high plains in the western part of the state. The North and South Platte join about 50 miles to the east of the Station to form the Platte River, which then continues inexorably across southern Nebraska to link with the "Big Muddy," the Missouri River, at Omaha. I have not seen the headwaters of the North Platte, which are about 50 miles to the east of Steamboat Springs in north-central Colorado. This branch of the Platte then flows almost due north to Casper in Wyoming, then south-east into Nebraska.

One Sunday in August of 2001, I wanted to see for myself the headwaters of South Platte (the Middle Fork), which are about 6 miles to the north of Alma, Colorado. I must say that this was one of the most wonderful drives I have ever taken. The trip began about halfway up Ute Pass from Colorado Springs, practically at the foot of Pike's Peak. Ute Pass is really not much of a pass these days, as a modern, four-lane highway now traverses it. The highest point of the pass is about 8500 foot, at a place called Divide, just a wide spot in the road, with a few bars, the Teller County jail, a couple of gas stations, and a grocery store. Continuing another 20 miles west, to the top of Wilkerson Pass, at about 9500 foot, brought me to one of the most spectacular sites in all of Colorado, South Park (Fig. 31). The early Spanish explorers referred to this place as Valle Salado (salt valley), because of the large numbers of salt springs that attract all sorts of wildlife. Later, French trappers used the Creole term, Bayou Salade, to describe it. The Mountain Men of the early nineteenth century bastardized the Creole name and called it Bayou Solado.

The Park includes a broad expanse of moraine flatlands, roughly 30 miles long and 15 miles from east to west. On the western side of the Park is a line of low-lying hog backs, called the Mosquito Range. The backdrop for these low-lying hills is an absolutely beautiful row of massive 14000+ foot, snow-covered peaks, e.g., Princeton, Harvard, Yale, Holy Cross, and Mount Elbert (at 14533 feet, the highest point in Colorado), called the University, or Collegiate, Range. By the time the Platte reaches Alma (Fig. 32), at the north end of the Park, it is still a first-order stream, about 4 foot in width and a foot deep, and the water is very, very cold! The river then wends its way south through the Park (Fig. 33). It is joined by

Figure 31   South Park, Park County, Colorado, about 40 miles north-west of Colorado Springs, Colorado. From this glacial moraine, substantial quantities of gold were removed in the last half of the nineteenth century. It is now mostly used for grazing cattle and raising hay.

Figure 32   The headwaters of the Middle Fork of the South Platte River near the old mining town of Alma, Colorado, north of South Park about 6 miles. There were several other mining camps in and around the Park, including Hartsel, Fairplay, and Buckskin Joe, among others.

Figure 33   The Middle Fork of the South Platte River as it wends its way through South Park.

Figure 34   The South Platte River in rugged Eleven-Mile Canyon about 15 miles east of South Park. Fly fishermen, campers, and hikers of all kinds and ages use the Canyon.

the South Fork of the Platte near the tiny community of Hartsel, about 15 miles south of Alma. At the southern end of the Park, a 150-foot-high concrete dam and Eleven-Mile Reservoir intercept the river. Outflow from the reservoir then enters Eleven-Mile Canyon. Like the Park, the Canyon is absolutely striking in its beauty (Fig. 34), with the fast-flowing South Platte as its centerpiece. The high and rugged granite walls seem to swallow up the many fly-fishermen, hikers, and campers who are attracted to this marvelous spot in the mountains. At the mouth of the Canyon is another dam and Lake George. Leaving this place, the river passes into the Tarryall Mountains, where the North Fork of the Platte joins it, before finally coming out on to the Great Plains at Denver.

   The Park is a true moraine, with extensive glacial deposits throughout. Prior to the Europeans arriving, the Park was a haven for wildlife, with buffalo (bison) being the dominant large species and antelope a close second. However, virtually every fur-bearing species of the Rocky Mountains were also present, ranging from bighorn sheep and grizzlies to prairie dogs. The large number of buffalo was highly attractive to the Ute Indians, a mountain tribe in that part of the Rockies. There were frequent incursions, however, by several of the nearby prairie tribes, e.g., Comanche, Kiowa, Cheyenne, and Arapaho, who sought the abundant game in the

Park. The result was a seemingly ageless series of usually violent confrontations between the Utes and their plains' rivals. With an abundance of beaver, the intrusion of the Mountain Men was inevitable in the early part of the nineteenth century, which meant even more conflict for the Ute tribe.

Gold was found in California in 1849, but the mountains of Colorado were largely ignored until the nationwide financial crisis of 1857. It was only then that a great many of the bankrupted easterners traveled west to Colorado in search of a new fortune and some, though not very many, found it. Gold was initially discovered in the Cherry and Chicago Creek areas near present-day Denver. By 1859, a true gold rush for Colorado was in full swing. "Pike's Peak or Bust" was frequently painted on the canvas covering the ox-drawn Conestogas as they labored across the Great Plains. When gold became scarce at the site of the first strikes, prospectors spread west into the mountains, including those of the South Park area. Inevitably, towns such as Alma, Fairplay, Buckskin Joe, Tarryall, and Hartsel sprung from nothing. The Hartsel community was first a trading post operated by Sam Hartsel who settled in the Park in 1862. From the original homestead of 160 acres, the Hartsel holdings expanded until they included some 200 000 acres. The ranch became a thriving cattle and horse operation, which dominated the Park for a period of time. Hartsel even built a hotel to take advantage of the hot springs present at the confluence of the Middle and South Forks of the Platte River. From 1860 to 1863, roughly $1 500 000 in gold was taken out of the ground in, and around, South Park, mostly through one-man operations. Because the Park was glacial moraine, the primary method of mining was the use of sluice boxes along the many creeks, and the South Platte itself. Other entrepreneurial types, e.g., traders, road builders, freighters, muleskinners, even railroaders, came into the Park to service the mining industry. All of these adventurers are gone now, and only Hartsel, Alma, and Fairplay survive, although there is really not much left in these once thriving towns. Today, the Park is mainly used for the grazing of cattle and the growing of hay. During my wanderings on that Sunday in August, I stopped by the roadside to take some photographs. It was eerily still, except for the whooshing sound made by the few cars that passed by, and the gentle rush of the South Platte as it meandered through the Park down from Alma toward Eleven-Mile Reservoir.

I should note here that another large river of similar importance to the ecology and biodiversity of the Great Plains has its origins not far from the

Middle Fork of the South Platte. In between the Mosquito and Collegiate Ranges is the Arkansas River. It begins near Turquoise Lake, not far from Leadville, about 40 miles west of Alma, as the crow flies. This river moves south along the western edge of South Park, then through the deep Royal Gorge, and out on to the prairies of southern Colorado near Pueblo.

One would think with all this water emerging from the mountains that the plains of eastern Colorado would be a veritable paradise, but this is not the case, and never has been. Despite Bates' reference to the "fruited plain" in *America the Beautiful*, these prairies are extremely harsh and quite arid, which is why General Zebulon Pike referred to this region of Colorado as the Great American Desert when he passed through during his early nineteenth-century exploration. However, when one approaches the two rivers out on the plains and accesses the wide valleys carved out by each of them, the land is lush with grass, and covered by stands of gentle cottonwood trees and weeping willows. In our younger days, my wife and I used to drive partway along the Arkansas River Valley to Colorado Springs from Wichita, or Newton, Kansas, on our way between home and college. After we were married and I was teaching at KBS in Michigan, almost every year we would drive the Platte River Valley across Nebraska on our way to, or from, Colorado. Both rivers are sometimes described as being a mile wide and an inch deep – only a slight exaggeration. I can remember one spring break taking a friend of mine home from school in Colorado Springs to Wichita. We crossed the Arkansas River at Dodge City in western Kansas. The bridge was almost a mile in length, but there was just a tiny flow of water in the river. He was from Nyack, New York, on the mighty Hudson, and he smugly remarked as we crossed the bridge, "Do you really call this a river?" As my luck would have it, 2 weeks later, after we had returned to school, there was a flood out in western Kansas at Dodge City, and the Arkansas was out of its banks by about 2 miles. I looked at my friend as we watched the disaster on TV, and responded to his earlier query, "Now you see why we call it a river." He got the message.

Before Europeans came west, the eastern part of the Great Plains was covered by buffalo grass, standing as high as 6 foot. Except for a couple of small national parks, this part of the grassland prairie is gone now, replaced by vast fields of corn, wheat, sorghum, and other grain crops. As one travels further and further west on the Great Plains, the grass becomes short, and scarce. Huge gullies and dry riverbeds mark the western landscape, many of them heading ultimately for the Arkansas or the Platte. Having grown up in south central Kansas, something that always struck me was the sense of enormous size and open space out on the Great Plains. I am convinced that the absence of a tree line exaggerates the feeling of expanse. Most of the "outsiders" who travel through this region want to leave it as soon as possible, probably because it seems to go on, and on, and on, and it does! But the natives of the Great Plains have great respect for this country. In part, I suppose this is because of its raw beauty and the feeling engendered from standing alone in the middle of a place where you can see the ends of the earth, or at least it seems. The Flat Earth Society would surely feel at home here.

The existence of the CPBS near the Keystone Reservoir on the North Platte River can be attributed in large part to the persistence of Brent B. Nickol and John Janovy, Jr. In fact, Brent told me in an interview for this essay that, "if John and I had not picked up that program, I do not think there would have been a field station at Cedar Point." A little background on these two interesting people is necessary as a way of introducing my first-hand experience at the Station in the summer of 2001. Brent told me that, as a second-year undergraduate student, he had picked up a book while browsing in the College of Wooster's bookstore. It was the fifth edition of Asa Chandler's *Introduction to Parasitology* (1955), and what he saw and read immediately absorbed him. He learned that Chandler's book was the text for a course in parasitology at the College of Wooster, taught by Ralph Bangham, and he immediately decided to enroll in it the following term. He was especially intrigued by the acanthocephalans, even though Chandler had devoted to this group just seven pages of more than 800 in his book! Brent's serious interest in these helminths has lasted throughout his entire career. While taking the parasitology course, he enlisted Bangham's help in doing an independent study, one of the things that attracted him to the College of Wooster in the first place. He was led by Bangham to the acanthocephalans of turtles and, in particular, to the genus *Neoechinorhynchus*. This was, and still is, an enigmatic group of worms. My students and I have spent some sweat on these same parasites and it is easy to see why Brent was so interested in them. At the time, the great acanthocephalan guru was Harley J. Van Cleave at the University of Illinois. According to Brent, "Van Cleave was of the opinion that there was but one valid species, *N. emydis*." However, at about the time Brent was consuming introductory parasitology, Van Cleave died, and Bangham suggested that Brent take a look at the group to see if Van Cleave was in fact correct about the genus being monotypic. The single problem with his independent study, as Brent related, was that "Bangham purchased only painted turtles from a supply house, and painted turtles do not have *N. emydis*, or any other acanthocephalan for that matter!" So, Brent spent the next 2 years looking for acanthocephalans in painted turtles when none was to be found. Naturally frustrated, when he decided he wanted to do graduate work in parasitology, he asked Bangham where he could find "a lot of parasites," and he was told, "go south." So, he ended up at Louisiana State University (LSU), where he worked first on amphibian acanthocephalans, then on acanthocephalans in woodpeckers and raptors under the direction of Harry Bennett who, coincidentally, was the last student of Henry Baldwin Ward. After completing his Ph.D. at LSU, Brent was immediately hired by the University of Nebraska, where he replaced the venerable Harold Manter who had just retired.

John Janovy, Jr. entered the University of Oklahoma in 1955, intending to swim for the varsity team and pursue a degree in civil engineering. He was successful in his first endeavor, even securing a scholarship in his second year. The engineering idea was discarded soon after entering the program because he found it to be "so boring." He proceeded to switch his major to mathematics, which he found much more challenging and interesting. Along the way, however, he also took several courses in the zoology department. In the spring of 1959, John and I first met in an ornithology course taught by the great ornithologist and wildlife painter, George Miksch Sutton. By this time, John had decided to obtain his Ph.D. in zoology. He approached Sutton with the idea of doing his Master's work with him, "but Sutton said he was not taking students at this level" and told him first to get his Master's degree, then come back. John ended up in the lab of Harley Brown working with the gymnostome ciliate, *Dipleptus* spp., a predatory protozoan that is actually capable of killing and eating planarians. After he obtained his Master's degree, he returned to Sutton who suggested that a dissertation on the birds of the Texas Panhandle would be a good problem. By this time, however, John was married, and he and his wife, Karen, were expecting their first child. He was deeply concerned about what to do. Moreover, by that time he had become interested in parasitology, in part because of his friendship with Dan Harlow, who was working with Professor Self on the parasites of pectoral sandpipers. John was encouraged to go and speak with Professor Self by Karen, who, by then, was working as Self's secretary. John finally decided to switch fields and took his first course in parasitology, a field course, at OUBS at Lake Texoma in the summer of 1962.

John was lucky, for when he made the change, Self, Dave Parmelee, and Vernon Scott had just secured a grant from the National Institutes of Health to work on what John called the B-V-P (an acronym for birds, viruses, and parasites) project. The idea of the study was to examine the movement and transmission of arborvirues and parasites in migratory birds. John became a research assistant on the grant and his Ph.D. research was to focus on *Plasmodium hexamerium* in meadowlarks and starlings. All of the fieldwork for his dissertation and the B-V-P project was done at the Cheyenne Bottoms, a large, but very shallow, body of water located in the middle of Kansas, near the small town of Hoisington. This shallow depression is part of a huge, underground aquifer that runs through this region of the Great Plains from north to south. The aquifer just happens to be at the surface in the Cheyenne Bottoms, and is an enormous attraction for all sorts of migratory birds on their way north into Canada in the spring and to the Gulf of Mexico in the fall. On finishing his Ph.D., and with Self's help, John, Karen, and their family moved east, where he went to work on a post-doc with Leslie Stauber at Rutgers University. It was with Stauber that John developed a research interest in leishmaniasis, one that was to continue for several years after moving to the University of Nebraska in 1966.

When John and Brent began going out to Cedar Point many years ago, I frankly wondered if they were going to find anything worthwhile. Having taught in Michigan at KBS, I was accustomed to a rich and abundant fauna of parasites and hosts, and lots of water. The Great Plains, I opined, could not possess much in the way of parasites, or a great diversity of hosts, and I knew it had very little water. But I forgot about the resourcefulness of parasites, as well as the resourcefulness of my friends, John and Brent. Within a short period of time, they were publishing fascinating accounts on all sorts of parasites. And John, in his own inimical and creative way, also carved out a wonderful niche for himself by writing a series of popular books describing his observations and experiences on the high plains of western Nebraska. I enjoy reading John's books because they are always so vivid, with their descriptions of people, places, and parasites, and his philosophical musings are a delight, even though I certainly do not agree with some of them! Two of my favorites are *Keith County Journal* (Janovy, 1978) and *On Becoming a Biologist* (Janovy, 1985), the latter I suppose because it describes so many parallels in my own career. In the first of these two books, John spends a great deal of time talking about the Platte River, and the valley that it has been carving for so many thousands of years. It was this river that greatly influenced his scientific work.

Several other things have always intrigued me about John and Brent. One of them was the enormous difference in their approach to both teaching and research. John was mostly a protozoan person, but I think I would consider him now as more of a generalist. As I described earlier, Brent works with the very specialized acanthocephalans. John is rather esoteric in some ways. He loves to philosophize. Brent is more of an empiricist. I asked Brent to describe their differences in teaching styles. "John," he related, "would say, let's find what we can, and then make up what we can." Brent, in contrast, said of himself, "I have in mind what we are going to try and do, or accomplish, much more rigidly than he does." Although I do not entirely disagree with Brent, having watched John teach his field parasitology course, I believe he knows exactly what he wants his students to find. My observation, however, is that he also wants his students to discover it for themselves. Despite their striking contrasts in styles and interests – perhaps because of it – a wonderful synergism developed between the two when they arrived at the University of Nebraska, and it has lasted to the present. The relationship spawned one of the finest graduate programs in parasitology in North America over the past 35 years.

John and Brent saw the need for a field station soon after arriving at the University of Nebraska in 1966. About 10 years later, a limnologist in their School of Biological Sciences, Gary Hergenrader, made them aware of a vacated Girl Scout Camp adjacent to Keystone Lake in western Nebraska. The Corps of Engineers created Lake MacConaughy, situated above Keystone, as a flood control and irrigation project in the 1940s. It is the largest body of water in the state and, according to Brent, "is located in the most biologically diverse area in Nebraska." Kingsley Dam, holding back Lake MacConaughy, is the second largest earthen dam in the world (Fig. 35). Lake Ogallala, immediately south of Lake MacConaughy, was the borrow pit for the dam. Lake Ogallala is very deep and the water is surprisingly cold, enough in fact to support a trout fishery, even though air temperatures frequently exceed 100°F during July and August of each summer. Ogallala, in turn, has a small diversion dam at its south end, through which irrigation water is released via two canals. The shallower, downstream part of this body of water is called Lake Keystone.

The creation of the Girl Scout Camp is a wonderful, but sad, commentary on financial generosity, volunteerism, jealousy, mismanagement, resentment, miscommunication, and, unfortunately, failure at the end.

Figure 35 Kingsley Dam on the North Platte River. This is the second largest earthen dam in the world. The large spout is water emerging under the dam from Lake MacConaughy at its south-west corner, just north of the Cedar Point Biological Station, Keith County, Nebraska.

Karen Janovy gave me several documents that relate the entire story. I wish I had the space to tell it, because it is a classic. In brief, it is about how two well-intended women in Ogallala, Nebraska, Mrs. Myrna Gainsforth and Mrs. Robert Goodall, attempted to construct a memorial to their husbands, did it successfully, then saw the project collapse. There is a moral in this unfortunate tale. It goes to the heart of how *not* to act when nonresident volunteers attempt to superimpose their will and rules on a local group, without careful respect being paid to wealthy donors who want to participate in the management of their project.

The original owners of the land were Burdett and Myrna Gainsforth. They had inherited the property from Burdett's parents who had home-steaded the land in 1911. After the Girl Scout fiasco, the camp was vacant for a period of time prior to the University of Nebraska's School of Biological Sciences undertaking negotiations with the Gainsforth family. John and Brent were both involved in this process from the outset. They thought they had reached a successful disposition in their negotiation, until the Gainsforth family learned that alcohol would be used at the field station, and then they balked. In fact, this almost stopped the deal from going for-ward. However, the family was promised that the alcohol would be used strictly for the preservation of specimens, and not for personal consumption. So, in the end, John and Brent got their field station, and the latter was appointed Acting Director.

With this problem resolved, and the Board of Regents convinced the field station was necessary for the University, the CPBS became a reality. In that first year, they were sorely pressed for time in arranging a full complement of courses for the following summer. They had just 3 months to decide what would be taught, assemble a faculty, make announcements regarding the program, recruit students, collect equipment, and do all the other things necessary to get a biological station up and running. But they succeeded and, in June of 1975, the first courses were offered at the new field station in western Nebraska. Brent taught helminthology and John handled protozoology the first year. In addition to teaching, the demands of running a brand new field station were considerable for Brent. By the next summer, Brent was appointed full-time Director and he made the decision not to do any teaching that summer. It was then that John was asked by Brent to offer a course in field parasitology, one that has been taught continuously since that second summer at Cedar Point. In 1979, John became Director, a post he held until 1986, and then again from 1993 to 1999. During these periods of time, the University of Nebraska purchased the land and the buildings from the Gainsforth Foundation.

Since John's field parasitology course has had such a long run, and I was aware that it is one of the few still being taught in North America, I decided that I needed to go there and discover for myself what was going on. So, in late July of 2001, I flew to Denver from North Carolina, then drove to our family cabins near Green Mountain Falls, north-west of Colorado Springs. The old cabins were built by my wife's grandfather about 90 years ago and have been extensively remodeled since then. While not winterized, they are none the less quite comfortable at about 8500 foot up in the Rocky Mountains. Colorado at this altitude is most pleasant from the standpoint of temperature (and in almost every other way), so I had plenty of time to catch my breath, I thought, for what was to come in south-western Nebraska the next week.

I have always enjoyed the part of Colorado east of the mountains, because it is so flat and you can see so far. It reminds me of Kansas where I grew up. There are very few trees, except those along the South Platte or Arkansas River Valleys. The drive to Cedar Point was uneventful. Indeed, I didn't even have to stop as the trip took only about 4 hours to cover the

250 miles or so. When I drove on to the station grounds in mid-afternoon, I spied John coming out of his lab. Even though I had been raised on the Great Plains, I was totally unprepared for the shock when I got out of the car. I am certain that everyone has had the experience of opening an oven door and having a blast of heat strike your face, and then of steam fogging your eyeglasses, if you wear them. Well, that was close to what I felt when I stepped out of my car and on to the Station grounds. Next to an oven, the nearest thing to south-western Nebraska at this time of the year would be a sauna. I greeted John and we went immediately to his lab where there was relief (it was air-conditioned!). He took me on a mini-tour and we then walked over to the cabin where his wife, Karen, warmly greeted me (she is one of my favorite parasitology wives in the entire world). We talked for a while about our families. Then, it was time to begin the more formal interview with John, which I taped for future reference. Essentially, I asked him many of the same questions that I had asked Brent a few weeks previously in Albuquerque so that I could get another perspective regarding parasitology at both University of Nebraska-Lincoln (UNL) and Cedar Point over the years. Karen listened in the background as we talked and occasionally corrected our sometimes faulty recollections whenever the conversation drifted to our joint experiences at the University of Oklahoma. Karen actually knew a lot more about a number of things than either John or I during those early days, especially since she had served as Professor Self's secretary for several years while John and I were graduate students in Norman. That evening, we all drove to the tiny community of Paxton where we enjoyed a wonderful dinner in a fascinating restaurant called Oley's Big Game Steak House. The place made you feel like you could be in the hinterlands of deepest Africa or the Arctic because of the huge variety of stuffed animals (not the customers) present in the dining room. All of them had been collected over the years by Oley, the long-time owner of the local eatery. By the time we drove back to the Station, the temperature had cooled considerably, probably to around 90°F by then. To top off the evening, there was a wonderful Great Plains sunset to admire as we headed west back to the Station.

The next morning, I was primed, or so I thought, for a day in the life of a field parasitology student, "Janovy style." I arrived at the lab in time to hear (yes, I said hear) the tail end of one of John's daily exams, about which I will talk a little more in just a bit. After the exam, John began to explain to his class what was going to happen during the rest of the day. One of the things that struck me was that he gave no really specific instructions.

In effect, he told his students that he wanted them to collect everything they could find that moved at the two field sites they were going to visit during the morning. He mentioned the names of a couple of different parasites they might encounter, but not much more. This matches closely to what Brent had told me to expect from John when I interviewed him, before the Nebraska trip. It also reflects on something that John said in our discussion on the Sunday of my arrival, to wit, "I look for problems, not for solutions."

I would be remiss if I did not relate one more story concerning John's teaching approach. Janine Caira, now at the University of Connecticut, obtained her Ph.D. at the University of Nebraska where she was a student of another marvelous teacher/scholar, Mary Lou Pritchard. Janine told me of her very first experience at Cedar Point during an American Society of Parasitologists dinner in Albuquerque in the summer of 2001. As she related it, she "had been accustomed to a somewhat more traditional style of teaching. The first summer after I arrived at Nebraska, I was excited because John had agreed that I could serve as the teaching assistant in his field course." She continued, "I remember the first day of class very well. Following a brief introduction to the intrigues of parasitology, we loaded the students into the field vehicles and drove them to one of John's favorite ponds, a site John knew to be home to leeches, deer flies, black flies, and, most conspicuously, mosquitoes. Upon our arrival at the pond, John nimbly sauntered out into knee-deep water and then encouraged the students to do the same, noting the numerous snails present in the pond. Once the class was in the pond, and sufficient time had elapsed for the various biting creatures to mount a significant attack on his class, John suggested that everyone count the number of mosquito bites they received over the course of the next minute. And, then, when the minute was up, John asked each student to count the number of bites he or she had received" in that single minute. After an appropriate pause, she related that John, "in his distinctive Oklahoma way," asked, "what does this tell you about the human condition?" Janine said, "With that one question, John put a whole new perspective on my concept of what constitutes good teaching." She ended her story by saying, "I knew that five weeks of *merely* [her emphasis and mine too!] trying to keep up was before me." In a similar way, I knew that *merely* trying to keep up on that hot summer day was going to be a real task as I trailed John and his class around the western prairies of Nebraska. Like Janine, I was certain it was going to be a great learning experience. I was not to be disappointed.

Figure 36   Neven's Pond on the stark prairie of western Nebraska.

On our way out to the first field site, one of the vans got stuck, in the dust, in the middle of a country road! This should illustrate how dry it can be out there – I cannot imagine what it would be like when that much dust turns to mud after a good thunderstorm. When we reached the first site, Nevens' Pond, I was really kind of surprised to find anything. As we drove into the pasture and parked, the only thing clearly visible was a windmill and what looked like an old cattle trough (Fig. 36). I should have known better, for leading away from the windmill and cattle trough was an overflow pipe carrying water to what could, I suppose, be called a pond. At least it was a shallow depression, and it was filled with water. The only problem was that it was also full of lots of submergent and emergent vegetation. John had instructed his students, "get everything," and that's what his class began to do, collect anything that moved, and even some things that didn't. At one point, I was standing beside John and he said, "look, see, the pond is full of bullfrogs" – my first tip-off that he knew exactly what was there and, moreover, exactly what he expected the students to find in the pond (Fig. 37). And it was indeed full of bullfrogs, as there must have been 30–40 of them sitting there with the tips of their heads and their eyes sticking out just above the surface of the weed-choked pond.

Something else that intrigued me about the pond was its setting. It was smack in the middle of a completely desolate prairie, with lots of dried brown grass, some scattered yucca, gently rolling hills, and just

Figure 37 Students in John Janovy, Jr.'s field parasitology class (summer, 2001), working in Neven's Pond.

one, lonely I suppose, but definitely forlorn, tree in sight. However, not far away were lush, green fields, many covered by acres and acres of feed corn and other crops. That is one of the interesting features about this small part of south-western Nebraska. It gives the impression of an almost desert-like sort of landscape, in remarkable combination with an abundance of what is the scarcest commodity of all, water. Everywhere you look, it is either very, very dry, or somewhat wet, the former because of the incessant sun and heat, wind, and lack of rain in this part of the Great Plains, and the latter from all the irrigation water transported into the fields from Lake MacConaughay. There was not a cloud to be seen that day. It reminded me of what we old baseball players used to call a "high sky." The only thing that saved me from the heat that morning was my straw hat and the gentle breeze blowing out of the south. It was then I remembered that the wind almost always blows on the prairies, either out of the south, or the north. Indeed, when my family and I arrived in North Carolina, I experienced an east wind for the first time – most shocking.

At any rate, the collecting went amazingly well. The students were unafraid to get wet or dirty! John was obviously pleased as they began picking up frogs, a couple of painted turtles, a wide range of aquatic insects, and even a few snails. Frankly, I was surprised by the diversity of critters the students managed to acquire in our relatively brief stay at the pond. They even found a 14-year-old box turtle wandering, apparently aimlessly, near the pond (they released it). I found it hard to believe that anything could survive out there. Moreover, if the air temperature was 100°F (and it was), the water temperature was easily 90°F (it had to be, at least). The students persevered and went about their business with enthusiasm and tenacity, while I observed, sweating away. Finally, John called a halt to the collecting, and everyone piled into his or her van and we left this extraordinary place.

I had driven my rental car into the field, knowing that its air-conditioning would bring at least a brief respite to the hostile environment. John had assigned one of his graduate students, Jaclyn Helt, to accompany me as we drove to the next site, I suppose so that I would not get lost inside the huge trail of dust created by the vans as I followed them down the country roads. I found Jaclyn to be a cheerful and bright young woman, and a very eager graduate student. She had already started her Master's research on the biology and life cycle of what she and John presumed to be a new species of *Phyllodistomum*, a gorgoderid fluke, which uses the killifish, *Fundulus sciadicus*, as the definitive host. She and John were certain the first intermediate hosts were fingernail clams, but they were not sure about the second host. My guess is that by the time this book is published they will either know for sure how the cycle goes, or at least have a very good idea.

We arrived at the second site in about 20 minutes, a place called Cedar Creek. Again, John's instructions were the same as the first time, "collect everything." By then, however, I knew exactly what he was doing, something I should have guessed earlier. As I mentioned previously, John's idea about teaching field parasitology, and about research in general, is, look for a problem first. Do not worry about the solution. In the case of the class field exercise, this was what John was trying to accomplish – find a problem. The students were being set up, although I expect that having been through 60% of the field course by that time, they knew where he was taking them.

At least Cedar Creek resembled a creek (Fig. 38). There was actually water moving through a few small channels, relatively cool spring water at that, and there was an abundance of live animals. The students again fanned out and, with an assortment of dip nets, seines, and enamel pans, began sorting through the debris they collected. Jaclyn and John did some seining for *F. sciadicus* and were also successful. The heat was growing worse as the morning passed, but the students and I were growing

Figure 38   Students in John Janovy Jr.'s field parasitology class (summer, 2001), working in Cedar Creek, Nebraska.

used to it by the time John announced we had accomplished our goals, and it was time to hit the road again. We next headed for a small, but very pleasant, wooded park in the tiny village of Paxton about 20 minutes away, the place where we had eaten dinner the evening before. We had a wonderful and relaxing lunch prepared by the Teaching Assistant for the course, Matt Bolek, who, I learned, was one of John's Ph.D. students, and was also an academic grandson of mine. It turns out that Matt was a Master's student of Jim Coggins, who had completed his Ph.D. with me at Wake Forest in 1975 and ultimately wound up as Professor at the University of Wisconsin-Milwaukee. It is a small world indeed, as I recalled that Jim Coggins had obtained his Master's degree with Jim McDaniel, an old friend of John's and mine, from our early days at the University of Oklahoma. Jim McDaniel had also done his graduate work with Professor Self, where we had become office mates and the best of friends. Unfortunately, Jim died way too young, and I still miss him. The lineage from Professor Self continues to grow as time goes by, a powerful indication of a strong mentor and the impressive graduate program he had established at the University of Oklahoma.

After lunch, we drove back to the Station and the lab. The teaching lab for all courses is in the basement of what they called the Lodge, the physical center of operations for the Station. During each 5-week summer session, three courses are taught 2 days each week, beginning on Monday and continuing through the following Saturday. Upstairs over the lab are the kitchen, dining, and library areas. The dorms, faculty quarters, and research labs are scattered mostly to the north and south of the Lodge. When we returned to the Station and unloaded everything, John gave us a break of 30 minutes to use as we please. I drove back to Ogallala for a quick shower and a change of clothes. By the time I returned, the lab was in full swing and I became the observer of a master teacher in action.

As we progressed through the afternoon, it was clear that John knew what his students would turn up at each site, but they did not at the outset, thereby invoking Janovy's First Rule, "seek a problem." For the students, the questions were, what parasites are present at each site, how do they survive, and what are their life cycles? If they could find the parasites, then a solution would follow. Another lesson, one that John had learned more than 42 years earlier, was being reinforced, for I am sure he began talking about it on the first day of class. Their lab manual, written by John, thus emphasized the need to record everything they did from the moment they stepped into the lab until they returned their microscopes to the appropriate cabinet and cleaned their tables from the day's dissections. To illustrate this imperative, John had included in his manual excerpts from a field notebook that he had kept in an ornithology course he had taken from George Sutton in the spring of 1959 at the University of Oklahoma. I know of its authenticity because I still have one of my own, written for the same course so many years ago. It is among my prized professional possessions, still!

While I observed John's teaching, I learned yet another lesson, one in pedagogy, that became important in my own parasitology course the next time I taught it. Most of us, when we want to demonstrate something in the lab, sit at a table and invite the students to gather round and watch as we make a dissection or perform some task. If we have three benches full of students, this means three separate demonstrations, each with the students drooping on their elbows and crowding against one another to see the "master" do his or her thing. What a waste of valuable time! John also does these sorts of demonstration dissections, but he does them with a set-up that includes two microscopes. Mounted on each, one a compound microscope and the other a stereoscope, is a Hitachi TV camera, both hooked up to a television set connected to a VCR. John not only shows all of the students the same dissection at the same time, via the TV screen, he records it on the VCR as well. The next class day, he uses the tape as a means of testing his students. The exam is thus based on the dissections witnessed by *all* of his students in the same series of demonstrations from the previous week. I was much impressed!

However, I think the thing that I really liked was the way in which John managed to tie everything together in such a coherent manner. He knew that at the first site they were going to find a species of *Haematoloechus* in the lungs of bullfrogs and metacercariae of the same parasite in the damselfly naiads. He hoped to find snails shedding cercariae so that he could show the students how the parasites are swept into the branchial baskets of the naiads where they then encyst. At the other site, he knew they would find a gorgoderid in the urinary bladder of the frogs, and he hoped to find cercariae being shed from the fingernail clams and metacercariae in the dragonfly naiads. He used the *F. sciadicus* he and Jaclyn had seined in Cedar Creek to show his students that another species in the same family of flukes used the urinary bladder as a site of infection, just like the gorgoderid in the frogs. First identify the problem, don't worry about the solution. Find the parasite first, and everything else will fall in place. If you know what you are doing, you will be able to lead your students where you want them to go, and you will be able to lead them into asking the questions they need to ask. What a great way to teach. This special skill is based on two things. First, there must be an intimate knowledge of the host–parasite system with which you are working. Second, there must be an equally great amount of knowledge about where this parasite is going to be found in the field, e.g., the pond has a lung fluke and the stream has the gorgoderid. Out on these prairies around Cedar Point, John has carefully honed both of these skills through many years of experience. In some ways, though, I believe his success is based on even more than this. While I was talking with him and Karen the next morning, I sensed something else that is of overriding importance in his teaching, and that is an utterly sincere pride in seeing a student succeed.

On a more personal note, as the afternoon passed, I wilted from the heat in the un-air-conditioned lab, despite the eight fans that were continuously whirling away. However, all they seemed to be doing was recirculate the super-heated Nebraska air. After supper and a seminar, I managed to slip away to the comfort of my air-conditioned motel in Ogallala.

The next morning I returned to find John and Karen in their apartment where I was invited to join them for coffee and a piece of pound cake. I accepted the coffee, but said no to the cake. I was sort of embarrassed when

Figure 39   John and Karen Janovy at Cedar Point Biological Station on the day of their 40th wedding anniversary (summer, 2001).

I learned it was the occasion of their 40th wedding anniversary and that the pound cake was a tradition on each anniversary (Fig. 39). I quickly reversed myself and accepted the offer. I thought it was the best pound cake I ever tasted! The rest of the morning was spent talking about places, mostly Cedar Point, and people, mostly John's students and those of Brent who had spent time at the Station since they began teaching there back in 1975. Many of these "kids" I knew. The numbers and the obvious talent that had come through the field parasitology program at Cedar Point were impressive. For example, John was able to identify 34 parasitology gradu-ate students who had done at least some of their thesis/dissertation work at the Station. Twenty-two undergraduates did research there. Seventeen in this latter group had come because of John's direct encouragement. Almost all of these undergrads were recruited out of John's introductory zoology course. I asked him how he managed to recruit so many. He sort of grinned and said, "well, at the beginning, I hand each of the 200–250 students in the course a 3 × 5 card and ask them to write a brief au-tobiography on it, including their interests in biology, if any. I then tell them to bring the card to my office. If they bring it, they will receive five points on their final grade in the course." This latter incentive induces about half the class to come to his office, where he then interviews each of them individually. John continued, "from this group, several have gone on to be very successful parasitology students," i.e., Mike Barger, Pete Olsen, Wes Shoop, Michele Carter. "You can tell immediately if they are articulate, if they are self-confident, etc. I ask if any of them are on scholarship." Of the 22 undergrads he has recruited in this way, nine have published papers in refereed parasitology journals and seven have presented papers at important scientific meetings. As John talked, I thought, not a bad record, indeed, not a bad record at all! And I began to see even more clearly why this veteran parasitologist and popular author had been so successful, and why colleagues like me really owe John such a debt of gratitude for enticing so many new and interesting folks into our professional ranks.

One of the things that we touched on that morning as we sat and talked about a whole range of things was how Professor Self had impacted our lives and those of his other students. I think we all three agreed that he was not so much a teacher as he was a giver of himself. As John said, the successful teachers and leaders of graduate students "are willing to give away some good ideas. I certainly see some people who won't do that, who won't give away their 'stuff,' who won't give away anything. I see some of our younger faculty who are too narrowly defined and won't give of themselves." I knew exactly what John meant. I too have seen young faculty members almost literally throw away their opportunities of creating a long-lasting graduate program in their departments by their early mishandling of graduate students. Bad reputations are quickly developed, and no matter how hard one tries to shake off such a reputation, it is almost impossible for a young faculty member to do it. A lot of this problem, I strongly feel, has to do with the way in which the young faculty members were treated as graduate students. Most of the time, we reflect the behavior and philosophy of our mentors. If the relationship is a good one, then the young faculty member will almost always be successful as well. In contrast, if the major professor is a "jerk," then that is the way young faculty members will deal with their students. It is unfortunate that the predilection for this kind of behavior is not something that can easily be picked up during a typical job interview.

John is not a one-dimensional faculty member. His mathematical training as an undergraduate gave him a huge "leg-up" with statistics and it is something he has used wisely and effectively over the years as a scientist and a teacher. I asked him what he considered to be his most significant set of research contributions and his answer was not at all unexpected. It was a piece of long-term work in the Platte River, something he started in 1975, the first year he taught at Cedar Point. As he described it, "that research was very revealing about the way the world operates. It is 'stuff' that tells me more about nature than is normally perceived." Although John and his students have published several papers dealing with this research, I think the most interesting summary of this particular work is based on a talk he gave to the Centennial University Professor Lecture Series at the University of Nebraska-Lincoln. It was published as part of the Nebraska Lectures (Janovy, 2000) and is entitled, *Who's Infected With Whom: The Natural History of Parasites* (I will note here that unless you see John is directly quoted from our conversation at Cedar Point, any other quotation comes from the lecture just identified). The reason I like the lecture and the paper that came from it so well is because, in this format, John was not constrained by the whims or biases of referees, associate editors, or editors. The lecture he had given and then subsequently published was "pure Janovy." Our discussion about it and his research was free-wheeling. Both venues gave him the opportunity to express his opinions and ideas without restriction. He was, in effect, teaching – something I have already pointed out that he does very well.

"That particular body of work," John told me, "really lays down for one system those factors that produce changes in numbers of parasites. It's as close as you can do it without doing an experiment." The key element in the study was his focus on the abiotic caprices of the Platte River, and the way in which these vagaries impacted the long-term dynamics in a community of parasites associated with *F. zebrinus*. The essence of John's thesis in this regard rests with two very fundamental principles, which, he asserted, affect or control the abundance of animal parasites and the richness of parasite communities. First, he contends "that time is a major confounding factor in studies on parasite abundance, mainly because transmission conditions change with time." The second "is that space equals time in this regard, in the sense that both impose environmental heterogeneity onto ontogenetic homogeneity. Translated, this principle states that the *conditions* under which parasite transmission must occur are variable and often changing, especially at the local level, whereas the *mechanisms* by which parasites must carry out transmission are evolved, thus fixed, at least for relatively long periods." An interesting thing, to me at least, about his long-term approach to this problem was that he recognized its significance "within two weeks of arriving at the station in the summer of 1975." He knew that this river, the plain's killifish, and its parasites, were going to occupy much of his research time each summer for a number of years.

Many parasites have seasonal life cycles, which means they are responsive to cyclical temperature changes. In this respect, temperature is not necessarily a vagary. It is usually an evolutionary construct to which the parasite's life history has been adjusted, thereby insuring that certain events necessary to complete the cycle will occur, on time, as it were. In the case of the Platte, changes in water temperature are both seasonal and predictable. So, although temperature is an abiotic variable, it is one to which the parasites of *F. zebrinus* have adjusted evolutionarily. The abiotic factor of overriding significance for the parasite community associated with the plain's killifish is, therefore, not temperature. It is water flow, which can "vary by at least an order of magnitude on annual, monthly, and almost daily basis, and further, the high flow rates can occur at various times of the year." In early July of 2002, for example, Ogallala was inundated by 9 inches of rain in just a few hours. Flooding of the South Platte River was so acute locally that the cascading water knocked out a bridge on Interstate 80!

Whereas flow rates vary, the life-cycle patterns of the parasites on a seasonal basis have been fixed through a long evolutionary process. In his study, John found that when flow rates increased, the prevalence and abundance of *Posthodiplostomum minimum* metacercariae in *F. zebrinus* decreased and that just the opposite would occur during periods of low flow. "In low water years, populations rise rapidly, a change related primarily to snail populations, which flourish in low water years, and in which parasite numbers are amplified by many orders of magnitude through asexual reproduction." High water flushes the snails from the local habitats where transmission of the cercariae from the snail to the killifish is most likely to occur. He also referred to *P. minimum*, correctly, as a generalist parasite. Glenn Hoffman, in his excellent book, *Parasites of North American Freshwater Fishes* (1998), reports that metacercariae of this parasite have been recorded from more than 100 species of fishes.

In contrast to the generalist, *P. minimum*, with its complex life cycle and ability to amplify its numbers asexually (via polyembryony) in the snail, is the specialist, *Salsuginus thalkeni*. This monogenetic fluke not only occupies a very narrow portion of the *F. zebrinus* gill surface, the plain's killifish is its only definitive host. The latter parasite never occurs in large numbers and has a direct life cycle. John points out, "over the long term most fish remain infected, but abundance remains low, varying only slightly, regardless of stream flow. In this case the worm's evolved life cycle does not include an amplification stage . . . and we can infer that most *S. thalkeni* transmission occurs in the quiet shallows where the killifish spawns."

One of the interesting conclusions drawn by Janovy from this elegant study relates to a hypothesis he generated based on the frequency distributions of these two flukes in association with the killifish. On the one hand, he found that *P. minimum* is highly aggregated in *F. zebrinus*, not unlike many other helminth parasites in an intermediate or a definitive host. On the other hand, *S. thalkeni*, while infecting a large number of killifish, is seldom found in high numbers, but is not aggregated in its distribution. In his lecture, he stated, "intuition and conventional logic suggest a strong founder effect, in which that successful fraction of the overall reproductive output must logically represent an equally miniscule fraction of the parasite species' genetic variation." He continued, "according to this logic, most worms should evolve rapidly, adapting quickly to new hosts, or responding quickly to host defenses." For *S. thalkeni*, "genetic variation is theoretically mapped broadly on host genetic variation, perhaps allowing the symbiotic participants to track one another through evolutionary time" – a most fascinating idea. However, as he concluded in his lecture, "testing a founder effect hypothesis using our generalist *Posthodiplostomum minimum* in *Fundulus zebrinus* has all the potential for becoming what might be best described as the Ph.D. problem from Hell."

As we continued to talk the morning of my departure from the Station, John went back to something he had told me about on the day of my arrival, and that was how he began the long-term work in the Platte. As he described it, a couple of ichthyology students came into his lab with several *F. zebrinus* and asked about the white spots on the gills. His graduate student, Steve Knight, and John determined that the white spots were *Myxobolus* sp. cysts. They began collecting more *F. zebrinus* from the Platte and counting the cysts. As they progressed with the enumeration, they began seeing other parasites in the gills and elsewhere on and in the fishes. He thought that someone ought to be following the entire parasite community, so when Ann Adams, a new graduate student, came into his lab, he placed her on the problem. By the time she had completed two more years of work on the killifishes in the Platte, John had done all the *Leishmania* research that he wanted to do and made the decision to spend the next 5 years working on the *F. zebrinus* parasite community in the Platte. After that 5 years was finished, he decided that he would continue the project and see where it would take him. As he described it, the study grew from just looking at *Myxobolus* sp. cysts, to examining the whole parasite community, to incorporating the various abiotic conditions in the Platte, all of which led to the *Who's Infected With Whom* lecture and a passel of other publications. I was fascinated by this tale because it parallels the one Clive Kennedy related to me regarding his long-term work in Slapton Ley, and it clearly follows the pattern for our studies in Charlie's Pond over the last 20 years. I think, in all three of these cases, that each of us began with a simple question, what was going on in the system with respect to a single parasite? As each of us, and our students, became more involved in resolving the initial question over time, we discovered other problems and questions, and the research grew to be something that none of us really intended at the outset.

I remarked to John that it seemed, philosophically, his teaching and research are cast in a very similar manner. He agreed, saying, "that's fair. I think it's an integrated approach and I don't have any clue where that idealism came from. You need to see all the parts, or as many as you can in order to see the way in which those parts work together." I asked him if, perhaps, this idea might have come from the B-V-P study in which he played such a critical role at the very beginning of his research career and he responded in the affirmative. But then he paused, and recalled his interactions with Professor George Sutton, the great ornithologist. He said, "to many people, a bird is just a source of tapeworms, but a bird to Sutton was an emotional thing, a career thing. It has a history, a geographical distribution. I do believe that Sutton's ideas regarding geographic distribution of birds at various taxonomic levels are a crucial part of the way I think about biology and the way I like to teach." He also included his undergraduate degree in mathematics as an important component in his teaching and research repertoire. He pointed out that, "it helps me to put a number on things." When he said that, I immediately recalled the response that Clive Kennedy gave when I asked him about the significance of Harry Crofton's 1971 papers regarding the dispersion of parasites. For me, John's comments were déjà vu, all over again. I could have pointed out that these ideas regarding the quantitative aspects of field parasitology actually extend back to Wisniewski's paper in the early 1950s, and even to Cort, McMullen, and Brackett, as early as 1937.

We ended the morning in a discussion about John's philosophy of doing science, and at about the same place when we began my visit 2 days before. Recall he said that it was his philosophy "to be a problem seeker rather than a problem solver." He told me that, about 15 years ago, he had served on the College of Architecture's reaccrededation committee. During a site visit, one of the outside members of the committee, a prominent architect from Washington, DC, made a formal presentation to the assembled students and faculty of the College. He said, "Do you want to be more influential, do you want to be more interesting, do you want to be better? You are already doing a superb job of training architects, but if you want to take the next step up, then you need to be problem seekers instead of problem solvers." He admonished his audience, "in your senior project, building a sky-scraper is a problem that has already been found. You need to go and find a problem that has not been solved." John does not just "talk the talk," he really does "walk the walk," when it comes to this idea. For example, whenever a new graduate student comes to his lab, he tells him/her to draft a list of 50 questions that he/she might like to do for their Master's or Ph.D. project. When the list is completed, he and the student sit down and go over it together. Each question is carefully considered and judged "as being implausible logistically, or of Nobel Prize quality but impossible to accomplish, or it has conceptual strength and would be real easy to do, etc." This was exactly the kind of thing that I watched him do in the field with his students and later in the lab. He forced his students to be problem seekers. He told them to find everything. Then, as he said to me, "we are going back to the lab and assemble our conceptual framework for what we have taken in the field." This is the way *he* works, and *it* works!

As I said in the previous essay, Will Cort and Lyle Thomas led the way in field parasitology, not only from being the first to teach it in North America, but the first to do a significant amount of research in this area. I am pleased I was able to go to Cedar Point, and to interview John (and Karen) and Brent about their teaching and research experiences. I also learned first-hand why John will be the recipient of the Clark P. Read Mentor Award from the American Society of Parasitologists in Halifax, Nova Scotia in 2003 (Brent won it in 2001). I honestly believe that Cort and Thomas would be happy with John's course. He is carrying their tradition forward in an admirable fashion, and I was fortunate in being able to see him do it for a while, in person!

References

Chandler, A. C. (1955). *Introduction to Parasitology*. New York, New York: John Wiley.
Cort, W. W., McMullen, D. B., and Brackett, S. (1937). Ecological studies on the cercariae in Stagnicola enarginata angulata (Sowerby) in the Douglas Lake region, Michigan. *Journal of Parasitology*, **23**, 504–32.

Hoffman, G. L. (1998). *Parasites of North American Freshwater Fishes*. Ithaca, New York: Comstock.

Janovy, J. J., Jr. (1978). *Keith County Journal*. New York, New York: St. Martin's Press.

Janovy, J. J., Jr. (1985). *On Becoming a Biologist*. Lincoln, Nebraska: University of Nebraska Press.

Janovy, J. J., Jr. (2000). *Who's Infected With Whom: The Natural History of Parasites*. Centennial University Professor Lecture Series. Lincoln, Nebraska: The University of Nebraska.

Wisniewski, W. L. (1958). Characterization of the parasitofauna of an eutrophic lake. *Acta Parasitologica*, **6**, 1–63.

Selected reading

Janovy, J. J., Jr. (1994). *Dunwoody Pond: Reflections on the High Plains Wetlands and the Cultivation of Naturalists*. New York, New York: St. Martin's Press.

Simmons, V. M. (1992). *Bayou Salado. The Story of South Park*. Boulder, Colorado: Fred Pruett Books.

6

The Canadians

The whole art of teaching is only the art of awakening the natural curiosity of the young mind for the purpose of satisfying it afterwards.

ANATOLE FRANCE, *THE CRIME OF SYLVESTRE BONNARD* (1881)

When I began thinking about this series of essays, I did not plan on doing anything with Canadian parasitology, even though I have developed many lasting friendships with some of the folks north of our border over the years. Indeed, three of my most unique Ph.D. students (Tim Goater, Dave Marcogliese, and Derek Zelmer) came down from Canada to do their work with me at Wake Forest. One of them, Dave Marcogliese, received the Henry Baldwin Ward Medal from the American Society of Parasitologists (ASP) in 2001. This prestigious award is given annually to an outstanding young parasitologist who has made a significant impact on research in his/her area of expertise. Tim was recently appointed to the Council of the ASP and Derek was just appointed as Associate Editor of the *American Midland Naturalist*. My mentor, Professor Self, used to say that he enjoyed basking in the reflected glory of his graduate students. Me too!

My reluctance in writing about the Canadians was not that a great deal of research in field parasitology had not been done there, far from it. It was mostly because I personally had done no work in the field up in that part of the world. Some of the time, and in most of the places, it is too damned cold (for me at least). Other times, there are too many blackflies, or mosquitoes, or both. However, several of the referees, and a couple of others, who read my book proposal asked, rather pointedly, why was I not going to include anything about Canadian field parasitology? Then, at the annual meeting of the ASP in Albuquerque in 2001, I was confronted by Tim Goater and David Cone, and challenged again. My arm was thoroughly twisted, almost off, and I succumbed to their urging. In agreeing to do it though, I first secured the promises of Tim and David, and Sherwin Desser, that they would supply me with some personal insights and background so that I could do justice to the essay. David Cone also loaned me a very informative book, *Parasites, People, and Progress. Historical Recollections*, written by the late Murray Fallis in 1993. In it, he describes the history of parasitology in Canada and some of what I will write about comes from that volume.

In doing the research for this essay, I decided that some of the focus would be on the work accomplished at the Wildlife Research Station (WRS) in Algonquin Park, located about 200 miles north of Toronto in Ontario. The real development of field parasitology in Canada, in point of fact, can be traced to the opening of the WRS in Algonquin Park in 1938. As noted by the late Murray Fallis in his book, "The stimulus for this particular research came from an interest in malaria and C. H. D. Clarke's description of a ... parasite, *Leucocytozoon bonasae*, found in the ruffed grouse." Considerable attention, especially by Fallis, was given to understanding the biology of *L. bonasae* and *L. simondi*, the latter a closely related species that is capable of causing heavy mortality in ducks. Much of the early effort at WRS was directed at the transmission of these parasites by blackflies, *Simulium* spp. According to Fallis, success with this work ultimately led to extended research on blackfly biology in Africa, Norway, and New Zealand.

A long line of Canadian investigators was to make many significant contributions to our understanding of wildlife parasitology at the WRS over the years. So, in late August of 2001, I sent an e-mail message to Roy C. Anderson and several others in this group, requesting some help with first-hand information regarding research at the WRS. Not 30 minutes after sending the message, I received a phone call from Sherwin Desser informing me that Roy had died from an apparent heart attack the night before. As I mentioned in the Prologue, his untimely and unexpected passing made me realize, all the more strongly, that the sort of biographical perspectives I wanted to share in this book were exceedingly important. Fortunately, Sherwin told me that about 3 months before Roy's passing, he had received a letter from Roy. In it was a detailed description of what he had done up in Algonquin Park, beginning with his interesting dissertation work on the filarial worm, *Ornithofilaria fallisensis*, and a mallard drake named Gus. Sherwin was kind enough to send this very detailed document to me, saying he "thought Roy would like you to share it with other parasitologists."

While reading Fallis' book, I learned that there was an active involvement with parasite problems in Canada going back more than 200 years. It seems that southern Ontario, for example, was greatly impacted by malaria during much of the eighteenth and nineteenth centuries. Fallis suggested that there were three possible sources for the malaria in what he referred to as "Upper Canada." The first was the movement into Canada of infected American and British soldiers during the Revolutionary War. Malaria was rampant in colonial America at the time. Indeed, as late as 1937, there were an estimated million cases of the disease extending even beyond the 42nd parallel in North America. Second, there was immigration of British loyalists (Tories) immediately following the revolution, many of whom were undoubtedly infected with the parasite. Third, during the mid nineteenth century, there was a huge migration of settlers from the UK to Canada. Most of us forget, or perhaps did not know, that malaria, also known as ague, had been a serious problem in England since well before the time of Shakespeare in the late sixteenth century. In fact, Chaucer mentioned both tertian fever and ague in his *Canterbury Tales*, written in the fourteenth century. In the *Nun's Priest's Tale*, he wrote, "You are so very choleric of complexion./Beware the mounting sun and all dejection,/Nor get yourself with sudden humours hot;/For if you do, I dare well lay a groat/That you shall have the **tertian fever's** [my emphasis] pain,/Or some ague that may well be your bane."

William George MacCallum, a Canadian physician who was trained at Johns Hopkins in Baltimore, was to make a significant contribution to understanding at least part of the malaria parasite's life cycle as early as 1897 when he examined the blood from a crow infected with *Haemoproteus* spp. Fallis wrote, "He [MacCallum] noticed, as others had for species of *Plasmodium*, the flagella (microgametes) breaking free from the parent cell (microgametocyte), wriggling among the blood cells, bumping into other cells and, finally, contacting and penetrating another cell (macrogamete)." This observation was made in the same year that Ronald Ross reported his important discovery regarding the significance of mosquitoes in transmitting the dread *Plasmodium* parasite, for which he won a Nobel Prize in 1902.

Interest for the support of general research in Canada was formalized with the establishment of the National Research Council in 1924. In 1928, responding to insightful businessmen, politicians, and academics, the provincial government of Ontario passed an act to provide $2 million for the creation of the Ontario Research Foundation, to be located in Toronto. The funding was to be used not only as an endowment for the Foundation, but as seed money to attract donations from private sources as well. Part of the research objective of the Foundation was "the mitigation and ... abolition of disease in animal and plant life and the destruction of insect and parasitic pests. ..." With the leadership of its first Director, H. B. Speakman, the Foundation was to build a strong and lasting relationship with the University of Toronto, one that was also to have an important impact on the development of parasitology, not only for Ontario, but also throughout Canada and, indeed, North America. Not long after the Foundation was created, the Institute of Parasitology, MacDonald College, McGill University, was established, and T. W. M. Cameron was appointed its first Director in 1932. According to Fallis, Cameron "built the Institute into the largest centre for research in parasitology in Canada and coordinated the work of a very skilled team of researchers."

The Ontario Research Foundation, primarily through the efforts of Murray Fallis, was to develop a close working relationship with the WRS in Algonquin Park over the years. The WRS attracted such parasitology "stars" as C. H. D. Clarke, Roy C. Anderson, Reino (Ray) Freeman, Gordon Bennett, Sherwin Desser, and their many students. One of the most productive in this group was Roy C. Anderson. As I mentioned above, Roy died just as I was attempting to reach him regarding an interview for this essay. I was greatly saddened by his passing, not only because we parasitologists lost such a productive scholar, but also because he was such a kind and decent man.

When Roy arrived at WRS in the early summer of 1951 to undertake research on his Ph.D. with Fallis, he was unsure about the kind of work he wanted to do. His roommate that first summer was Gordon Bennett, who was later to have a distinguished career working on nasal bots and blackflies in Maylasia, Australia, and then at Memorial University in Newfoundland, before dying unexpectedly on Christmas eve, 1995. After toying with the idea of work on filarial worms in beaver and porcupines, Roy finally settled on a filarial worm parasitizing ducks. As he said in the document sent to Sherwin Desser, he "finally decided to return to my first love, namely birds." This was an excellent choice, for a couple of reasons. First, Murray Fallis was in the process of looking at the biology of *Leucocytoon simondi* in Algonquin Park ducks, and Roy had discovered microfilariae in the blood of those ducks that survived the vicious attack of this apicomplexan parasite. Second, the life cycles of avian filarial worms were unknown, even though Sir Patrick Manson had observed microfilariae in the blood of crows as early as 1878.

As I read Roy's account of his study on this filarial parasite, I became exceedingly impressed, first with his systematic approach to the problem, and then with his exceptionally keen observational skills. The very first thing he did was to locate and describe the parasite, the easy part. It was a new species, which he named *Orinthofilaria fallisensis*, in honor of Dr. Fallis. He had a good laugh on himself with respect to the species name, however. He noted, "the ending *ensis* was, of course, a silly error that no one noticed until it was published"; in Latin, *fallisensis* means "from a place called Fallis." No matter, he had the parasite identified and knew what it looked like. But the real detective work remained. How was it transmitted? His attempts to infect mosquitoes were unsuccessful. He observed, however, that ducklings maintained outside the lab in cages would become infected and, moreover, microfilaremias always developed in May and June, the peak in blackfly activity in the park. The suspected vector was *Simulium venustum,* and he was on track of resolving the problem. When he fed uninfected *S. venustum* on an infected duck, he was successful in following and describing the development of the parasite in the blackfly. He then isolated the parasites from the head parts of infected flies and inoculated them back into uninfected ducklings. After 30–60 days, these ducklings developed microfilaremias and he was subsequently able to isolate the adult filarial worms under the birds' skins. As he put it, "I had fulfilled Koch's postulates and discovered the kind of vector transmitting the parasite."

This is where Gus, the duck, became part of Roy's story. He wanted to know if the microfilaremia in the duck was periodic or not. It should have been since *S. venustum* is a daytime feeder. So, he took the bird back to his third-floor apartment in Toronto and placed it in his bathtub where it was to stay for 7 straight days (Roy never said how he and his wife managed to bathe during this time). Every 4 hours for 7 days, he was to take a "measured sample of blood from a leg vein" and count the number of microfiliariae. They were periodic, confirming his hypothesis. He described Gus as a "sociable creature who hated to be alone and he quacked all day and night as only a frustrated drake can do." But, he and Gus made it through the ordeal, despite the great consternation of his wife and their neighbors, as Gus quacked constantly for 7 days and nights. Roy continued that "Gus recovered his composure quickly from the experiment," and he returned the bird to Algonquin Park the next spring.

Unfortunately, Gus died, in Roy's lap, even as he was performing a feeding experiment with blackflies. Roy greatly lamented losing his prized duck and his sentiments regarding the bird were aptly stated, "The data poor Gus provided immortalized him in the *Canadian Journal of Zoology*."

It should be noted at this point that, although *S. venustum* was shown to vector *O. fallisensis* successfully, it was not the vector in nature. It also does not vector *L. simondi*, which puzzled investigators, including Murray Fallis, for many years. The problem was that the blackflies, which feed on waterfowl, were not *S. venustum*, but *S. rugglesi*, a discovery made by G. E. Shewell of Agriculture Canada. The former species was in fact a "human biter" and the latter was found to feed exclusively on waterfowl. Subsequently, they were to discover that the ornithophilic *S. rugglesi* would feed only on ducks that were sitting on water. If ducks were placed on shore, a few feet from the lake, they would not become infected, or if they were placed over water, but a few feet above it, they also would not become infected. Anderson stated that the misidentification problem set back studies on leucocytozoonosis at least 25 years. He wrote, "It's an excellent example of the need for taxonomy and taxonomists . . . we need these people now more than we need additional molecular biologists who know little natural history and who are blind to the real problems in nature. Any molecular biologist interested in zoology should first complete a degree in systematics and then turn to molecular methods. He or she would not be at a loss for problems to investigate." Not that I have any disrespect for the molecular folks, but I would add, touché (!), to Roy's comments. I strongly feel that molecular biology is a formidable cornerstone for our discipline, but, without the classical systematist, parasitology would be "dead in the water." If we do not know the identity of the beast with which we are working, how can we say anything in a definitive manner about its molecular or phylogenetic relationships to other organisms, or to other biological systems for that matter? It would be like having salt without pepper, truly a dismal thought!

One of the remarkable aspects of Roy Anderson's career at Algonquin Park was his capacity first to recognize a problem (sounds like John Janovy, Jr.), then investigate it, and always thoroughly. He saw, for example, the unusual nematode, *Diplotriaena* sp., was in fact not a parasite of the body cavity of birds as had been previously described in the literature. Instead, it was present in the air sacs of its avian definitive host. This resolved the question of how eggs of these parasites exit the host, a problem about which he had mused for a long period of time (because the literature asserted they were parasites of the body cavity). He found that eggs from gravid females are carried from the air sacs by ciliary action and mucus to the back of the throat (Roy called this the "bronchial escalator") where they were then swallowed and shed in the birds' feces.

Dictophyme renale, another nematode to which he became attached, is the giant kidney worm of mink. An individual female of this species can grow to a length of 25–30 cm. Remarkably, these worms always occur in the right kidney and up to 40% of the mink in the park are infected. Inside the kidney, the functional tissue is eliminated and replaced by a huge cyst with a villous lining that secretes lymph in which the parasite is bathed. Interestingly, the ureter leading from the infected kidney remains open and eggs are readily shed to the outside via the ureter and urinary bladder. Roy had begun work on the life cycle of the mink kidney worm, but he was to be "scooped" by E. M. Karmanova, a parasitologist in the former USSR. Karmanova claimed that an oligochaete, *Lumbriculus variegatus,* was the intermediate host, something that Anderson confirmed. Later, however, two of Roy's graduate students, Tom Mace and Lena Measures, were to discover that centrarchid fishes and frogs were paratenic hosts for the parasite and it was via these piscine and amphibian hosts that the parasites actually infected the mink. Recall that paratenic hosts are inserted in a number of parasite life cycles as a means for bridging trophic gaps. This explained to Anderson the outbreaks of the kidney worm in Canadian sled dogs when fed fish in enzootic areas. Humans can even become infected with the giant kidney worm if they eat infected and undercooked fish.

On reading the autobiographical sketch regarding his research in the park, I believe Roy took the greatest delight in his work with the so-called meningeal worm that occurs in white-tailed deer. The park was full of deer, which were frequently struck and killed by motorists while passing through. It was during the necropsy of several of these deer that a meningeal worm was discovered in the cranium. When first seen in deer, it was thought to be a filarial worm because of its long, slender shape. Roy soon learned that it was, instead, a metastrongyle. Eventually, he decided that the parasite should be described and its life cycle determined, and he accomplished both tasks. The taxonomic problems associated with the worm were quite perplexing initially. Ultimately, the parasite's name was assigned to *Parelaphostrongylus tenuis,* the identity of which was based on the description of a single male worm from the lung of a deer in eastern North America several years previous to the initiation of his study. The life cycle was another matter. His first step was to isolate larvae of the parasite from the feces of wild deer, which he did. Then, because of the parasite's taxonomic position, he surmised that the larvae were being transferred to deer via small terrestrial snails as the deer grazed. So, he collected snails and slugs in areas of the park where deer were not found and exposed them to larvae isolated from deer feces. The larvae penetrated the snails and developed to the L3 stages. His guess was correct. The next step was to find uninfected deer, preferably fawns, as they would be easier to handle in the experiments he wanted to attempt. It took him a while, but he eventually located a source, ran the experimental infections, and was completely successful.

A truly fascinating aspect of this work, however, was the manner in which the eggs from females in the cranial spaces were being shed as L1 larvae in the deer feces. He related that in one of his experiments he found 10 gravid females and five males in the cranium of an experimentally infected fawn 115 days postinfection. These females had laid their eggs on the dura mater and some had even hatched. However, and most importantly, they (he and his assistant, Dave Gibson) "found several females in the intercavenous sinus, a broad expansion of a vein near the pituitary. In this location, eggs would pass down the venous system to the jugular and right heart. From there they would be carried with the blood to the lungs via the pulmonary arteries. We took pieces of fixed lung tissue and sectioned it [them] for examination with a compound microscope and found exactly what we expected. Eggs were caught up in the fine capillaries between the air spaces. They became surrounded by inflammatory cells, which formed small nodules. Inside the nodules, the eggs embryonated into first-stage larvae which hatched, broke out of the nodules, entered the air spaces, the bronchioles, bronchi and trachea and, aided by the flow of mucus, moved by ciliated cells to the back of the throat," i.e., by means of his so-called bronchial escalator. From there, they were passed through the gastrointestinal tract and to the outside in the feces. Roy and Dave had completed the entire cycle of this marvelous little nematode and what a brilliant piece of work it was! Further research on the pathology associated with the parasite in white-tailed deer revealed that it was minimal and that the host's immune system was in some way culpable in maintaining the parasite at reduced numbers, which, in part, kept the trauma of infection in deer at a very low level.

Roy's long-term interests in filarial worms led him to the work of a Japanese investigator, Chuzaburo Shoho, who had made an interesting observation during World War II on the differential effects of *Setaria*

Figure 40   A young female moose infected with *Parelaphostrongylus tenuis*. This animal was weak in the hind quarters and was unable to rise to a standing position without help. (Courtesy of Murray Lankester.)

digitata, a filarial worm in cattle and horses in the Far East. In the body cavity of cattle the parasites were benign, but when passed to horses, they damaged the central nervous system and "produced paralytic disease." It was a classic example of a parasite behaving quite innocuously in its normal definitive host, but "causing a serious clinical disease" in an abnormal host. This fortuitous observation by Shoho was to lead Roy into an investigation of what was known as moose disease (Fig. 40), and yet another brilliant discovery. In Algonquin Park, and elsewhere in certain areas of North America, moose were known to suffer from a peculiar neurological problem which produced lethargy, ataxia, paraplegia, and, ultimately, death. I recently had the opportunity of viewing a most effective video created by Annie K. Prestwood and Murray Lankester, entitled *Parelaphostongylosis.* In it, they showed several moose with the disease and it is a terrible thing to see these huge, lumbering beasts with absolutely no control over their ability to walk, or even stand, without falling over clumsily. At any rate, when Roy coupled his discovery of the *P. tenuis* life cycle and the strange behavior of *S. digitata* in two different kinds of hosts, he thought that perhaps the meningeal worms in deer could be causing problems in moose. According to Anderson, "It had been speculated that the disease was caused by ticks, tick-transmitted bacteria, or cobalt deficiency."

But, to him, these explanations seemed implausible because of the "basic features and distributions of the disease in moose." Like many an epidemiologist in search of a disease-producing organism, he obtained all the reports he could find of moose paralysis and plotted them on a map. He discovered that moose paralysis occurred only in the southern regions of the animals' distribution and only in eastern North America. There were two clues, however, that emerged which placed him on the real track to discovering the cause of moose disease. First, moose had been studied on Isle Royal in Lake Superior for many years, but there had been no reports of paralytic animals on the island. Second, Newfoundland had an expanding and healthy moose population and, again, no moose paralysis. The one thing these two islands had in common was the absence of white-tailed deer!

With the help of Murray Lankester, then an undergraduate student at MacMaster University, they acquired two moose calves (Fig. 41) and had them shipped to WRS. They obtained *P. tenuis* larvae from infected snails and placed them in milk, which was bottle-fed to the calves. Then, they waited. In the meantime, Shoho had arrived from Japan on a visit to the WRS. The story regarding Shoho's arrival at the Toronto airport and subsequent trip to the lab is amusing, although somewhat embarrassing for

Figure 41 A young Roy C. Anderson and Murray W. Lankester, with one of the two calves used to confirm that *Parelaphostrongylus tenuis* from white-tailed deer would cause moose disease. (Courtesy of Don Robinson.)

Roy. After picking up his guest at the airport, they traveled back to the lab. When they arrived at the gate, Roy discovered he had misplaced his keys to the lock on the chain blocking the road into the lab. It was apparently very hot and humid that day. Just off the plane, Shoho was still dressed in his business suit, complete with tie, and they had no way of getting inside the gate. So, they had to wait for someone with keys to the lock. To make matters even worse, his car was not air-conditioned. Roy described their torment. "Two hours passed. No one was in sight. Nothing stirred anywhere except the mosquitoes which invaded the car and danced on the windows and attacked us. We slowly melted." He continued, "My guest was clearly gloomy. He looked straight ahead at the chain wondering, I imagined, how he'd got himself into such an intolerable situation. Then Shoho broke the silence and asked, 'Doctor, I wonder you don't take chain off nail.'" His visiting colleague was correct! The chain was hanging on a nail on one side and the padlock in the middle was a sham, "a fact obvious to anyone who took the trouble to examine it with any care. What is worse, I realized immediately that Shoho was in possession of that crucial fact from the time we were first confronted by that damn chain and I have often wondered what was going through his mind as he observed me behaving in such an irrational manner. I got to know Shoho well over the years and I knew there was no point in asking him about it. He would have been too polite to say."

When they finally got inside and dried off, cooled down, and away from the mosquitoes, Shoho examined the moose calves and then told Anderson and Lankester to be patient, the neurologic signs and symptoms would eventually appear. He was correct. Within 3 weeks postinfection, the animals were suffering from full-blown cases of "moose disease." At necropsy, *P. tenuis* adults were discovered in the cranium and larvae in the lungs. As importantly, there was severe damage to the dorsal horns of the gray matter in the spinal cord. Anderson and Lankester had conclusively demonstrated that they could experimentally induce moose paralysis with *P. tenuis*. In Roy's mind, however, one more step was necessary and that was to link the parasite in the brain of a moose directly with the paralytic disease. Anderson and Lankester obtained the heads and spines of three moose with neurologic signs of the disease and the infection by *P. tenuis* was diagnosed in all three animals. Anderson was to say, "Thus, we had proved conclusively that the cause of moose sickness was a neurotropic nematode transferred from deer to moose where the ranges of the two cervids overlapped in North America."

Roy wrote in his autobiographical sketch, "My discovery of the cause of moose sickness when others failed over a period of years taught me one thing about research. Except for accidents one is most likely to find something if one has some idea what one is looking for. That was a strong factor in my success." As he pointed out, others had come across these worms in moose, even in the brain stem, but they were looking for something else. "Also," he said, "how could they suspect that these slender insignificant nematodes could cripple a large animal like a moose?" How indeed! After leaving the Ontario Research Foundation and going to the University of Guelph in 1965, he continued thinking about the moose paralysis problem. He eventually concluded that "the main reason moose encountered *P. tenuis* in recent times was the expansion of deer range northward in the past 100 years as a result of environmental changes brought about by humans. Forestry practices, fires, and agricultural activities created suitable deer habitat and they multiplied in areas inhabited by moose," thereby accounting for the spread of *P. tenuis* and its benign relationship with one cervid to the highly pathogenic and lethal relationship of the parasite in another cervid. What an interesting man, and what an interesting body of research in field parasitology this man generated in 14 years at the park and the WRS. In 1997, Roy received the first Clark P. Read Mentor Award from the ASP. The next year, it was Austin J. MacInnis' turn, and I received it in 1999. I am pleased to note here that Brent Nickol won the award in 2001 and John Janovy, Jr.will receive it in 2003, quite an eclectic group of recipients. As I look at the late Roy Anderson's great career, I cannot think of a better person to have received it the first time. I am even more honored to have received the same award knowing now so much more about his work as a teacher and a researcher. In my mind, he epitomized all the characteristics of a true scholar.

I was to develop a number of close relationships with Canadian parasitologists, but my earliest began a little more than 35 years ago. In the summer of 1967, I invited Reino (Ray) Freeman to visit the W. K. Kellogg Biological Station (KBS) and present a seminar. After receiving his Ph.D. at the University of Minnesota in 1952, Ray was first employed by the Ontario Research Foundation, with a cross-appointment in the School of Hygiene at the University of Toronto. I had been working on aspects of the biology of the bass tapeworm, *Proteocephalus ambloplitis*, for a couple of years in Gull Lake at KBS and was anxious to share my findings with him. Needless to say, I was more than somewhat chagrined to learn that not only had he and his graduate student, Hertwig Fischer, been working on the life cycle of *P. ambloplitis* in Lake Opeongo in Algonquin Park, they had also determined that the published accounts of its life cycle were inaccurate. As it turned out, I had been following the wrong trail. When Ray told me what was going on with this tapeworm, the data that I had assembled from my work in Gull Lake, however, suddenly made good sense. The old literature had indicated that the definitive hosts were primarily large- and smallmouth bass, that copepods were first intermediate hosts (for procercoids), and that fingerling fishes were second intermediate hosts (for plerocercoids). However, as determined by Fischer and Freeman, the cycle was actually much more complicated. For example, one of the observations that I had been making was that juvenile smallmouth bass were not infected with adult tapeworms, and I can distinctly remember Ray asking me about this when he visited KBS that summer. It was then that he told me the whole story regarding the so-called bass tapeworm and its life cycle.

Fingerling fishes acquire the parasite when they feed on infected copepods. The procercoids migrate into parenteric sites and become plerocercoids. Sexually mature definitive hosts acquire the parasite when they prey on infected fingerlings. However, these plerocercoids do not directly develop into adult tapeworms; instead they migrate into parenteric sites and accumulate there until the following spring. Then, according to Fischer and Freeman, when water temperatures rise above 65°F, parenteric plerocercoids migrate into the intestine and become sexually mature. Adult tapeworms are lost in the fall and the process is repeated the next spring with a new cohort of parenteric plerocercoids. If the parasite enters the tissues of a bluegill sunfish, or most other centrarchid species, the plerocercoid remains in the parenteric site and does not respond to the temperature stimulus. The fact that the plerocercoids migrate from parenteric sites is not questioned. However, the idea that temperature is the sole stimulus is in some dispute. Thus, all fishes are impacted by the same seasonal temperature changes, but parenteric plerocercoids migrate into the intestine only in their appropriate definitive hosts and only in sexually mature individuals. Herman Eure, one of my former graduate students here at Wake Forest (but now Professor and Chairman of Biology at Wake), worked on the bass tapeworm in Par Pond, the large cooling reservoir on the Savannah River Plant about 1000 miles south of Algonquin Park. He found that adult *P. ambloplitis* were present in largemouth bass during the winter and not in summer – additional evidence that a spring rise in temperature is not the sole stimulus for migration. Unfortunately, however, we are still confounded by the actual stimulus for plerocercoid migration.

Ray's work up in Algonquin Park extended to a number of other parasite taxa, including several cestodes. One of his legacies that still impacts parasitology in many parts of the world, some 40 years after the work was completed and published, was his research on the cestode, *Taenia crassiceps*. This is a truly remarkable parasite. The life cycle includes canines, primarily foxes in nature, as the definitive hosts. The cysticercus of the parasite is normally found in subcutaneous tissues of microtine rodents. The unusual feature of this larval stage is that it buds exogenously and each tiny daughter bud will develop into a complete new cysticercus, which, in turn, will produce additional buds. Ray found that if he removed larval *T. crassiceps* (the ORF strain, named for the Ontario Research Foundation) from a rodent, and placed the cysticerci and tiny buds into Petri dishes with physiological saline, he could then draw several small buds into a syringe and inoculate them into the abdominal cavity of a laboratory mouse. He would then wait for a few months and kill his artificially infected mouse. On necropsy, the abdominal cavity would be full of cysticerci and buds. In effect, he had substituted the syringe for the canine definitive host. Most importantly though, the inoculated mouse would eventually be in possession of enormous quantities of cestode tissue. In other labs, *T. crassiceps* has been maintained in this fashion for many years and used in a wide variety of important physiological, biochemical, and immunological studies. Unfortunately, for me personally, and for all parasitologists, Ray Freeman developed Alzheimer's disease and died young, way before his time as far as I am concerned.

Another of the talented Canadians who spent a great deal of time in the field was the late Leo Margolis (Fig. 42). Leo was a Ph.D. student of T. W. M. Cameron at the Institute of Parasitology at McGill University in Toronto. After completing his degree in 1952, he headed west to the Pacific Biological Station at Nanaimo, British Columbia, where he made several highly significant contributions to understanding the biology of marine parasites. While undertaking many administrative and other responsibilities during his long career at the Nanaimo station, there were two research areas in which he particularly distinguished himself, at least in my opinion. The first involved the biology of *Anisakis* spp. This is a nematode parasite that uses marine mammals as its definitive host, but can also cause anisakiasis, a nonlethal but excruciatingly painful disease in humans. After hatching in the open sea and molting once, the L2 stages

Figure 42 The late Leo Margolis, wearing the Order of Canada medal, given for his many significant contributions to research involving the salmon industry in North America. (Photographer unknown, but photograph obtained from Bob Kabata and Tim Goater.)

of the larvae are consumed by microcrustaceans where they develop to the L3 stage, after molting again. If either a macrocrustacean or a fish eats the microcrustacean, the parasite will enter the tissues, but remain an L3. The macrocrustacean, or fish, is a paratenic host, because it bridges an ecological, or trophic, gap in the parasite's life cycle. The marine mammal is thus unlikely to eat a microcrustacean, so the parasite employs the second host, which is much more likely to be eaten, as a way of gaining access to the fish-eating definitive host. Consumption of raw, or partially cooked, fish is the way in which humans, instead of marine mammals, acquire the stomach worm. The importance of Margolis' work on anisakiasis had strong economic implications for the herring industry throughout the world. The larval stages of *Anisakis* in the flesh of canned herring are rather disgusting to see, although since they are killed in the canning process, they are not harmful to anyone who might consume them in this form. He found, however, that larger fishes possess more of them, and that the worms are also larger. So, his advice to the industry was use small herring for canning. This resolved the problem.

A second significant contribution by Leo was his development of ways to increase salmon production in the north-western USA, Canada, and Alaska. There were several problems in the management of these fisheries and Leo cleverly addressed them. For example, it is important to distinguish between salmon stocks originating in various areas, to understand their oceanic migratory routes, to know breeding and feeding areas, and to understand the recruitment patterns of juveniles into stocks. Margolis' solution was simple. Use parasites as tags rather than conventional tagging methods, which are both labor-intensive and not totally conclusive. He was thus able to distinguish sockeye salmon from North America because of the presence of plerocercoids of the cestode, *Triaenophorus crassus*, whereas those from across the Bering Sea possessed the nematode *Dacnitis truttae*, which was acquired in the Kamchatka River in Siberia. Using these two parasites as markers, Margolis found that salmon were able to disperse some 2000 km from their spawning streams, and that they overlapped in their oceanic distribution. Even so, he showed that schools of North American sockeye tended to remain distinct while at sea. Some of the schooling salmon, however, could be separated from each other by the differential prevalences of *T. crassus* plerocercoids, which were characteristic of their spawning stream or river in North America. The procedures developed by Margolis and his esteemed colleague, Bob Kabata, at the Pacific Biological Station in Nanaimo, have been used by a number of investigators to answer similar questions for other species of fishes. In addition, they have been applied to other types of biogeographical problems involving marine parasites in the various seas and oceans of the world.

Whereas Leo was a world-class parasitologist, he had a diversity of interests and pursued many activities away from his professional responsibilities. The following story came from David Cone and it illustrates a little known facet in Leo's other life. It seems that Leo had become interested in ice hockey as a young man and played while he was in college. The love of this sport continued when he moved to Vancouver after receiving his Ph.D. He became closely linked to the British Columbia Amateur Hockey Association, and was its President from 1963 to 1966. He was to receive the 1990 Order of Merit Award from the Canadian Amateur

Hockey Association as a way of acknowledging his contribution to the sport over the years. These accolades for both his scientific and community involvement led him to receive an honorary degree from St. Mary's University in Halifax, Nova Scotia in 1992. Leo was to say that "he had been totally taken by surprise; he simply had not received any prior hints that this was going on." David relates that at the pre-convocation dinner, he and his wife, Lynn, were seated at the same table as Leo and his wife, and Archbishop Burke, who was to introduce Leo the next day. At dinner, the Archbishop presented Leo with a card. On the picture side of the card was a sepia-toned shot of Burke in action as a goalie, with his thin leather glove stretched in the process of making a spectacular save. On the back of the card were his statistics, impressive ones at that, for the 2 years that Burke had played in goal for the St. Anne team. Burke had tried out for the Montreal Canadiens in 1932, unsuccessfully, and later that year entered the seminary, ice hockey's loss but the Catholic Church's gain. In David's note to me, he intoned that Leo had written him a wonderful letter describing his great joy on that marvelous occasion.

Sadly, Leo died at age 69 from complications following cardiac arrest. It is my understanding that Leo suffered the fatal attack while walking home just before his retirement from the Station – a real tragedy for Canadian, US, and international parasitology. For all of his efforts, he received a great many official recognitions and citations, in addition to the honorary degree from St. Mary's University. For example, in 1975, he was elected as a Fellow in the Royal Society of Canada and was elevated to an Officer of the Royal Society of Canada in 1990. He was recognized by the ASP in 1995 with the Distinguished Service Award and, in the same year, with a Gold Medal from the Professional Insitute of Public Service of Canada (the Government Union) for outstanding contributions to pure and applied science. Finally, my friend Tim Goater tells me that "Leo Margolis was the recipient of Canada's greatest honor, the Order of Canada, for his outstanding scientific achievement, especially his tireless international work with the Pacific Fisheries Commission dealing with salmon stock management."

Although he is now retired, another of the legendary Canadian field parasitologists was based, and still is, on the Atlantic side of the continent, at the University of New Brunswick. I am, of course, referring to Mick Burt. And, when I say he is retired, this must be considered in a totally different context as compared to the majority of us regular mortals. Among many other things, for example, in his "retirement" he recently served as

Chairman of the Tenth International Congress of Parasitology's (ICOPA X) Scientific Committee! Even while performing this monumental task, he took the time to share with me some personal insights into his career, and descriptions of a few of his moments as a field parasitologist, for which I am greatly appreciative.

A native Scot, Mick obtained an undergraduate Honours degree from St. Andrews University in Scotland. During our exchange of correspondence, I was curious and asked him if he had played any golf at the famous St. Andrews course during his school days. Seemingly with great pride, he replied in the affirmative. In fact, from the way he responded, I believe he was actually addicted to the sport in his early years. That is easy to understand considering the mystique of golf in Scotland, and at St. Andrews in particular. I learned that he was a member of the New Golf Club in St. Andrews and even had a season ticket. He told me that when he first started playing in 1951, a season ticket cost about 10 shillings (= $2.00!), and when he left in 1961, it was about £5 ($15.00!). Moreover, the season ticket gave him access to any of the four courses in St. Andrews. The only caveat was that tee times had to be booked for the Old Course and Sunday play was not permitted. On completing his undergraduate studies, he left to pursue his Ph.D., but came back to complete work on it at St. Andrews in 1962. On his return, he said, "I played even more golf with a fellow parasitologist post-grad, C. E. Fitches, going out every morning during the week in the summer months. We would start at 6:30 am, be back for breakfast at 8:30, and in the lab by 9:00 am. Two hours was standard time for a twosome with no one in front of you." He then added, "I do not play now because: (a) it is ridiculously expensive; (b) it takes too long (over 4 hours usually); and (c) because of (b), I lost interest. I also lost my ball (!), and gave my clubs to my son."

Mick spent the summer of 1962 at the Institute of Parasitology at McGill in Montreal, where he started his Ph.D with the distinguished T. W. M. Cameron, before returning to St. Andrews in the fall. It seems that he could not support his wife and child on the Canadian graduate student stipend of $1340/year. He said, "With scholarships in the UK totaling over 2000 pounds sterling (about $5000 equivalent), this was not difficult as the cost of living was *much* lower than in Canada. For these two years, I was supervised by my father," who, I might add, was also a well-respected parasitologist of that era.

Mick's "migration/transposition to the New World actually started in 1957–1958 where," as he said, "I got a taste of North America (corn on the cob, hamburgers and hot dogs, ice hockey, fall colours)." For this brief period, he was a student at Union College in Schenectady, New York. Then, after a year, he returned to St. Andrews. However, even before completing his undergraduate studies at St. Andrews, he was drawn back to the New World when he was invited to the University of New Brunswick (UNB) for a 1-year appointment as an assistant professor in 1961–62. He had apparently made a real impression on A. R. A. Taylor, who was a Professor at UNB and on leave at St. Andrews at the time. "UNB needed someone to teach Invertebrate Zoology and General Biology (I had majored in Botany and Zoology as an undergraduate), because of the move of E. Hagmeier from UNB to Victoria and the delay of his replacement." Mick continued, "I enjoyed UNB and they seemed happy with what I had done as I was invited back in 1964. I had finished my Ph.D. research and 'all' that needed to be done was to write it up. This took me three years instead of three months."

Mick met his present wife, Barbara MacKinnon, in 1975. Barbara took her Ph.D. in parasitology with Alan Pike at the University of Aberdeen. They were married in 1987 and are the proud parents of two daughters. He described their relationship as follows: "How much we both loved parasitology was apparent when we were discussing how and where to spend our summer holidays. We both wrote down what we would like to do and where we would like to go. Although the locale differed (I proposed Scotland; Barbara proposed Australia), the activity we wrote down was collecting new and interesting parasites, elucidating their life cycles, and finding out all we could about their reproductive biology and host/parasite relationships." If this is not first-order dedication to field parasitology, I do not know what it is!

Coincidentally, Mick has also had a long association with another St. Andrews, this one in New Brunswick. It was there, in St. Andrews, during 1969, that a consortium of universities and government agencies in eastern Canada came together to establish what was first called the Huntsman Marine Laboratory (the name was later changed to the Huntsman Marine Science Center, or HMSC). This new lab was appropriately named for A. G. Huntsman, who had been Professor at the University of Toronto and a long-time Director of the federal marine research lab at St. Andrews. The federal lab eventually became the St. Andrews Biological Station, and land adjacent to the station was then acquired from the Federal Department of Fisheries to create the HMSC. According to Mick, the new marine lab was "to use this land in order to establish a year-round teaching and research facility which could be used by university scientists and their students enabling them to collaborate and interact with the staff at the St. Andrews Biological Station."

Mick began using the facilities at HMSC in 1970 and continued to use them for the remainder of his career. His research interests were varied. They ranged from turbellarians (*Urastoma cyprinae*) in oysters and other marine bivalves, to trichodinids that supposedly jump from shellfish to flounders. He was involved with developing a way of using the parasites of herring as bioindicators of stock identity. A major contribution was made by Mick and several co-workers to understanding the biology, life cycle, systematics, in vitro culture, and intermediate host status of *Pseudoterranova decipiens*, one of several species of sealworm that cause anisakiasis. In our correspondence, Mick's list of the many scientists with whom he collaborated over his long and distinguished career is impressive. I will not attempt to include them here because there were 44 names.

One of the interesting things I learned about Mick was his ability to combine the teaching of marine parasitology with inspired culinary creativity. The acquisition of this special talent was undoubtedly stimulated by an admonition from his father, the late Professor D. R. R. Burt who, as I mentioned earlier, was a well-respected parasitologist in his own right. His father had said, "never waste an animal by killing it only for its parasites." Mick continued, "This translated into eating the hosts, once they were cured of their parasites, and the development of many novel and tasty recipes such as periwinkle pizza, garlic sea cucumbers, curried seal, and mustard herring, to name but a few." Dave Marcogliese told me that one summer when he was conducting research at the HMSC, where Mick was teaching, he and Fred Purton (an HMSC technician) were asked by Mick to collect a seal for his class to necropsy. No doubt, Dave thought, he wanted his students to see some of the ubiquitous anisakid sealworms. So, Dave and Fred headed out to the bay and shot a harbor seal on a nearby island. They returned the seal to the lab. Dave went on to relate that Mick's class was about to engage in what could be called the "esoteric and Epicurean" portion of his parasitology course. It seems that each student is required to cook and eat the host he or she had selected for the class necropsy. In addition to the Atlantic cod, flatfishes, sea urchins, and sea cucumbers, Dave's seal ended up in a very special curry, superbly prepared by Mick of course. Dave did not say if the seal had any anisakid worms.

I asked Mick for a representative recipe that I could share in this chapter and he gave me the one for garlic sea cucumbers. By the way, Mick said to me that, "Reprints and recipes are available to serious-minded, fun-loving, parasitologists who engage in field work and who enjoy what they do." What follows is quoted almost verbatim from a Mick Burt e-mail:

Garlic sea cucumbers (Ingredients: sea cucumbers, garlic, ginger, butter, lemon juice, sea salt, pepper)

Modus operandi (Instructions)

1. Slit open fresh sea cucumbers and remove the guts for examination for parasites (mesozoans, parasitic turbellarians, nematodes, others).
2. Using one of your own thumbnails (or the back of a teaspoon), separate the pink muscle basket from the thick outer "skin" (which is white on the inside).
3. Wash both the muscle basket and the skin in seawater or salted water (3–4% saline) and preserve separately.
4. For the muscle basket, which will roll up like a crumpet, slice transversely into thin (about 1-cm) strips and marinate in fresh lemon juice for a few minutes with a sprinkling of salt and pepper.
5. At medium heat, melt enough butter to cover the bottom of a frying pan and add freshly sliced ginger root. Simmer for 2 or 3 minutes (don't let the butter burn). Add whole cloves of garlic (one/sea cucumber) and simmer for another 2–3 minutes (don't let the butter burn).
6. Pour the sliced sea cucumber with its marinade into the pan and simmer for a further 2–3 minutes (don't let the butter burn).
7. Serve as an appetizer, allowing one sea cucumber per person. (Enjoy!)
8. For the skins, place them in a large pot and boil in salt water for 1 hour. Decant the fluid down the sink and replenish with cold salt water. Boil for a second hour and throw away the fluid. Repeat for a third hour.
9. Slice the tenderized skins into strips (about 1 cm thick) and marinate as for the muscle basket in lemon juice, salt and pepper, or in Teriyaki sauce.
10. Either cook as for the muscle basket, or fast-fry with onions in olive oil, to which a few drops of sesame oil have been added. Serve as a side dish with fish, lobster, or any seafood of your choice. Excellent with rice!
11. Note: Both of these are beautifully complemented with a Sauvignon blanc wine.

For Mick, field parasitology was not confined to the marine habitat, but has involved parasites in freshwater systems as well. On occasion, he and his wife, Barbara, would travel to Hay Bay on Lake Ontario, during the summers for their holiday. As indicated previously, they consistently made every effort to combine pleasure with parasitology. On these excursions to Hay Bay, they were often accompanied by visiting parasitologists with whom they shared their cottage and the excitement of productive interaction with parasites in the local fauna. On one of these trips, he and Barbara hosted Lena Jarecka, a Research Associate at UNB. They had previously discovered that the local bowfins (*Amia calva*) were infected with the truly enigmatic cestode, *Haplobothrium globuliforme*, a parasite with which I had also had some early experience at Gull Lake during my KBS days. I remember removing a bunch from a necropsied bowfin and watching them in a fingerbowl under a dissecting scope when, suddenly, out popped four spined tentacles from the scolex of one of the worms. I was not only startled, I was certain that freshwater fish should not have cestodes with spinose tentacles. I quickly referred to my Wardle and McCleod and discovered, to my dismay, that the parasite had been described by A. R. Cooper way back in 1914 in the *Transactions of the Royal Society of Canada*. In addition to a tentacled scolex, the tapeworm exhibits one of the most unusual developmental characteristics with which I am familiar. This remarkable developmental feature involves sexually undeveloped proglottids, which break off from a primary strobila (on which sits the scolex with the spinous tentacles). Each new proglottid then develops a pseudophyllidean-like scolex at one end. The other end of the proglottid then undergoes strobilization, with each new proglottid in the secondary strobila developing pseudophyllidean-like reproductive systems. It really is a weird developmental process, without parallel as far as I know among other cestodes.

The life cycle of the parasite was not known at the time, providing the team of Mick, Barbara, and Lena Jarecka a good summer challenge. They had some insight into the biology and eating habits of the bowfin and, knowing of the cestodes' pseudophyllidean morphological characteristics, they had some ideas regarding its life cycle as well. They had watched the parasite's eggs hatch and observed the ciliated coracidium's swimming behavior. Toward the end of their stay, they collected large numbers of copepods, the suspected first intermediate host. They placed the copepods into well-rinsed brown beer bottles to which they added eggs of the cestode. As Mick tells the story, by that time they were ready to return to UNB. "Air Canada was most cooperative, once they confirmed that the open beer bottles contained water from Lake Ontario and that this was harmless to their crew or other passengers, and treated us with respect and reverence that eccentric academics clearly deserve." Mick continued, "We were given many amused looks by fellow passengers on Air Canada as Lena and Barbara each sat side-by-side with a 12-pack of open beer bottles on their laps all the way from Toronto to Fredericton." (Mick sat several rows behind Barbara and Lena and pretended not to know them.) When they got back to UNB and examined the copepods, they discovered that a number were infected with procercoids. They studied the parasites by light and electron microscopy, eventually publishing their results in support of Janicki's Cercomer Theory.

They had also collected 101 brown bullheads and necropsied them the night before their departure from Hay Bay. Of the first 97 they examined, none was infected. They were about to give up, even though they still had four remaining (Mick thought they had collected just 100, but had miscounted and thus the odd number). It was nearly 2:00 am, they were tired, and they were departing for home (with their brown beer bottles) later in the day. Then, they hit pay dirt, as "#98 had three plerocercoids, #99 had two plerocercoids, #100 was clean and #101 had five plerocercoids. I'm not sure what this says for standard statistics: going from 0% prevalence in a sample of 97 to 75% in a sample of four! All the fish were caught at the same site, in the same net, at the same time." I agree with Mick's assessment regarding the statistics. However, with some additional sampling, the data also sound like there may be a strong case for overdispersion.

On another trip to Hay Bay, Mick and Barbara were accompanied by Bogdan Czaplinski (from the Polish Academy of Science and then President of the World Federation of Parasitologists), Rod Bray (of the Natural History Museum in London), Lena Jarecka, and Bob Shephard. Their accommodation consisted of a small cottage, with two small bedrooms, each with a small double bed. Providing bed space for everyone was obviously going to be a problem. Mick tells the story. "After much discussion and coin-tossing, we agreed that the two ladies (Lena and Barbara) would share one bed in one room and the two international guests (Bogdan and Rod) would share the other bed in the other room. We also agreed that Bob Shephard and I would each have the couch for half the night and the easy chair for the other half. Having lost the coin toss, I was assigned the couch from midnight until 4:00 a.m. and Bob had the chair. Whether it was the close proximity of unfamiliar bodies that contributed to the insomnia of those in the bedrooms with the luxury of a real bed was never determined. I am adamantly sure, however, that their lack of ability to sleep had nothing to do with the gentle, sonorous murmurs (totally consistent with a clear conscience, a good meal and adequate libation) that I was accused of making."

Life in the field brings other problems on occasion. Mick related that "the excitement and fun of collecting and working there [Hay Bay] was increased by the necessity of finding alternatives to standard laboratory equipment [as Plato said, "necessity is the mother of invention" – GWE's interpretation]. A shower stall with several bags of ice served as our walk-in refrigerator. A jug of tap water with a handful of salt provided excellent saline (about 1%) for cutting mucus and cleaning worms prior to fixing. Although more expensive, when all of the [normal] fixative had been used up (or spilt), hot vodka or gin make good alternatives. The kitchen table (when it rained) or the park table/bench were just as good as our lab benches back home and sunlight was far superior to small light bulbs, especially during examination using the dissecting microscope." In concluding his correspondence, he gave me a real sense of his research philosophy. He said, "On all of our field trips, the well-known adage purportedly coined by Confucius (the great Chinese philosopher who lived in the first century BC) had special significance: Hear, and forget; See, and remember; Do, and understand!"

Harry Crofton's two papers, published in 1971, were classics with respect to parasite population ecology and epidemiology. On the community side, there are three papers written by John Holmes (Fig. 43) for the *Journal of Parasitology*, back in the early 1960s. I believe most investigators would agree that these three publications were as important to the development of parasite community ecology as those by Crofton were to parasite population biology. An interesting aspect of the Holmes' papers, however, is that they were all based on a series of experiments conducted in the laboratory, yet their major influence has clearly been on field parasitology. Although John's earliest roots (Minnesota) make him a Yank, he has worked in Canada throughout his entire career, teaching and doing research at the University of Alberta in Edmonton, or at the Pacific Marine Station in Nanaimo, British Columbia. He started his Ph.D. at Rice University with one of parasitology's really great scientists, Asa Chandler. When Chandler died, he came under the tutelage of the renowned Clark Read. The thrust of his dissertation research was actually quite basic – probably one reason that he got so much mileage out of it. Simply stated, he demonstrated the effects of interspecific competition between an acanthocephalan, *Moniliformis dubius*, and a cestode, *Hymenolepis diminuta*, in the gut of a rat. On the surface, this may not seem like very much, but what it did

Figure 43 John Holmes, Tim Goater, and Cam Goater (left to right) in the summer of 1994, just before departing on a cruise with their students. They were teaching Ecology of Marine Parasites at the Bamfield Marine Station on the rugged coast of Nanaimo Island, British Columbia, Canada. (Photographer unknown, but photograph obtained from Tim Goater.)

was to stimulate about 40 years of work, discussion, and debate on the way in which parasite infracommunities are organized. John has continued throughout his career to provide enormous insight and contributions in the arena of parasite community ecology.

Not long ago, the *Journal of Parasitology* began running a series entitled *Defining the Field*. The idea was to reprint a classic paper from the *Journal* and then invite an expert to create a follow-up by providing a thoughtful discussion on the current status of the work covered by the classic paper. At the beginning of this series, it was solely my decision as to which paper would be reprinted and who would write the follow-up (John Oaks and I are doing the assigning together now). Holmes' early papers on competition certainly met the criteria necessary regarding their status as classic and I decided the first one of the three should be part of the series. However, I was unsure who should write the companion paper. So, at the ASP meeting in San Juan, Puerto Rico, in 2000, I asked Holmes to make a suggestion. His reply was, "Let me think about it overnight." The next morning, I found him and asked if he had cogitated on who should write it. His response, without hesitation, was, "Yes, John Janovy." So, I asked Janovy if he would do it and he agreed. His paper appeared in the *Journal of Parasitology* in June of 2002. For someone with an interest in parasite community ecology, I found this review of Holmes' work to be one of the most well-written, informative, and provocative papers that I have read. Because of the high regard in which I hold this work, I will not deal with Holmes' contributions any further here, but instead refer you to the Janovy paper. He has captured the essence of Holmes' research so well, and that which followed, there is no way I can improve on it.

John Holmes, in addition to being an exceptional scientist, has also been a very successful mentor. Among those who did their Ph.D. work with John and then continued to contribute significantly over the years is another Yank transplant, old "red-beard" himself, Albert O. (Al) Bush (Fig. 44). Among other things, Al is one of my favorite people and closest friends. I first met Al, and his wonderful wife, Maggie, about 20 years ago, even before he completed his Ph.D. degree. Al did his Master's research with Don Forrester at the University of Florida before heading for Alberta and John Holmes. When he completed his dissertation, he took a job at Brandon University in Brandon, Manitoba. One of the things that I have always admired about this man is his total commitment to teaching. At

Figure 44   A group of parasitologists enjoying a brew on a sunny afternoon in August, 1988, just before the annual meeting of the American Society of Parasitologists in Winston Salem, North Carolina. Left to right: me, Al Shostak, Al Bush, Tim Goater, and Dave Marcogliese. (Photographer unknown, but photograph obtained from Tim Goater.)

Brandon, his load is heavy, but he loves every bit of it! From all that I have heard from several of his former students, among whom include Tim and Cam Goater, and Dale Edwards, he not only loves teaching, he is also very good at it. Al made his initial research mark with his dissertation on parasite communities in lesser scaup ducks. This study was significant for several reasons. Probably the least important result, but none the less impressive, was the number of parasites he found in these birds. (I was going to say "recovered," but Al is adamant that recovered is the wrong terminology to use in situations like this one, so I will defer to him in this case.) In the 45 birds shot at several Alberta lakes, there were some 1 million enteric helminths, representing 52 species! Personally, I cannot imagine counting this many worms, let alone identifying all of them. More importantly, however, was Al's successful attempt to define these parasites in terms of their guild structure, using terminology that had originally been developed for free-living species by R. B. Root some 20 years before. In the same papers, subsequently published with John Holmes in the *Canadian Journal of Zoology* in 1986, Bush also introduced the idea of parasite infracommunities, providing a parallel to the infrapopulation concept developed some 10 years earlier. And, as if this was not enough, he also applied the core/satellite concept of H. Caswell and I. Hanski for the first time to the helminth infracommunities, yet another idea that had been originally developed based on work with free-living organisms. All of these applications have seen wide use by parasite community ecologists following those publications in 1986.

Another thing that I have admired about Al's record over the years has been his collaborative efforts with field parasitologists who have worked variously with parasites in fishes, amphibians, and birds, as well as in freshwater, marine, and terrestrial habitats. He has been a contributor to several books, both as a co-editor and as a co-author. All of these efforts have had a strong ecological theme and have been well received by those working in field parasitology. I have had the good fortune of working directly with Al on several of these projects and it has always been an invaluable experience for me personally.

One of the most interesting of the Canadian field parasitologists is another transplanted Yank, Dan Brooks. I have known Dan for many years and consider him to be a very good friend. After doing his undergraduate work and Master's degrees at the University of Nebraska, he went down to the Gulf Coast Lab where he completed his Ph.D. with Bob Overstreet. His interest in parasitology was spiked by Brent Nickol when he took Brent's introductory course at Nebraska as an undergraduate. After some postdoctoral work at Notre Dame, he became a faculty member at the University of British Columbia, then moved to the University of Toronto in 1989. Along the way, in his wonderful career, he also managed to secure the H. B. Ward Medal from the ASP. You may not always agree with Dan, but almost anyone would concur that he is a consummate systematist, one of the best in the world. He is absolutely in love with cladistics, phylogenetics, and evolutionary biology. Although I have not seen his CV, my guess is that he has actually done little fieldwork in Canada.

A man of varied parasitological interests, he has recently been caught up in a project that sounds, at least to me, pretty "far out." He is associated with a group of world-class systematists and the All Species Foundation, who have as their goal to describe all of the biota in the world! One of the field sites for this monumental endeavor is the Guanacaste Conservation Area (ACG) in Costa Rica, a conservation region that extends from the Pacific to the Atlantic. Covering some 1100 km^2, it includes a wide range of habitats, ranging from mangrove swamps to cloud forests, and is home to some 235 000 species of plants and animals (exclusive of microbes), or about 2.4% of the total number of species on the planet. The goal of Dan and his group of parasitologists is to document all of the eukaryotic parasite species among the estimated 940 species of vertebrate animals in the ACG, and they are well on their way, having necropsied more than 2 000 animals so far. I mentioned in Chapter 3 that Derek Zelmer, a former graduate student of mine, and Dan had described *Halipegus eschi* in a frog from the ACG in Costa Rica.

Dan is a man of enormous energy and enthusiasm, which, on occasion, have landed him in the middle of a controversial issue of one sort or another. This propensity for controversy – fortunately for those of us with an interest in evolutionary biology and parasitism – has never deterred his appetite for discovery. If I had to choose anyone to lead a parasitology project of the scope envisioned by the All Species Foundation for the ACG in Costa Rica, it would have to be Dan Brooks. His commitment to the study of biodiversity and development of biological inventories is unrivaled.

It is difficult, in many ways, to separate field parasitology in Canada and the USA. There has been so much exchange of people and ideas that distinctions are blurred and often difficult to make. Many Americans have gone north of the border and assumed permanent residence at Canadian universities where they have made considerable contributions to field parasitology. Four of them are (were) close friends of mine. Ray Freeman did his Ph.D. at the University of Minnesota, then worked at the Ontario Research Foundation and the University of Toronto throughout his great career. I have already mentioned John Holmes and Rice University with Asa Chandler and Clark Read. Al Bush was connected with Holmes at the University of Alberta before going to Brandon University. Dan Brooks is yet another of these transplants, presently serving at the University of Toronto. There are (were) others who had similar career experiences, I am sure. Likewise, many Canadians have traveled south, four of them into my own lab here at Wake Forest. Tim Goater came down from Brandon to work with me and is now successfully ensconced at Malaspina University-College in Nanaimo, British Columbia. (I should note here that although he did not work with me toward any of his degrees, Tim's brother, Cam, and I became friends when we met in 1987. At the time, he was a Ph.D. student in Clive Kennedy's lab at the University of Exeter – he was also an undergraduate student with Al Bush at Brandon!) Dave Marcogliese (Fig. 44) did his Master's degree at Dalhousie University in Halifax before completing his Ph.D. here at Wake Forest; he is now with Environment Canada in Montreal and continues to do significant research in field parasitology, especially along the St. Lawrence River and its environs. Derek Zelmer worked at the University of Calgary for his Master's degree, did his Ph.D. with me, and is currently Assistant Professor at Emporia State University in Kansas, and Associate Editor for the *American Midland Naturalist*. Al Shostak (Fig. 44) did his Ph.D. with Terry Dick at the University of Manitoba, then spent 2 years with me here at Wake Forest, supported by a National Science and Engineering Research Council (NSERC) fellowship from the Canadian government. He is now at the University of Alberta and Associate Editor for the *Journal of Parasitolgy*. All five of these young Canadians are among the best young field parasitologists in North America.

One of the points I am trying to make is that the personal relationships and training of many Canadians and Americans are closely linked, and even reciprocal in many ways. Canadian field parasitologists have made a number of unique and very important contributions in their own right over the years, and I have made an effort in this chapter to tell a few of their stories. Finally, in formally recognizing some of the Canadian contributions to field parasitology, I hope that I have satisfied the supplications of David Cone and Tim Goater. I hesitate to tell either one so openly, but I am glad that I finally gave in to their forceful ministrations because I learned a lot about this hardy race, residing on the frigid side of the 49th parallel in North America! And, finally, how aboot trying some of Mick Burt's garlic sea cucumbers one of these days, eh?

References

Bush, A. O. and Holmes, J. C. (1986). Intestinal helminths of lesser scaup ducks: an interactive community. *Canadian Journal of Zoology*, **64**, 142–52.

Crofton, H. D. (1971a). A quantitative approach to parasitism. *Parasitology*, **62**, 179–93.

Crofton, H. D. (1971b). A model for host–parasite relationships. *Parasitology*, **63**, 343–64.

Fallis, A. M. (1993). *Parasites, People, and Progress. Historical Recollections*. Toronto, Ontario, Canada: Wall and Emerson.

Holmes, J. C. (1961). Effects of concurrent infections on *Hymenolepis diminuta* (Cestoda) and *Moniliformis dubius* (Acanthocephala). I. General effects and comparison with crowding. *Journal of Parasitology*, **47**, 209–16.

Holmes, J. C. (1962a). Effects of concurrent infections on *Hymenolepis diminuta* (Cestoda) and *Moniliformis dubius* (Acathocephala). II. Effects on growth. *Journal of Parasitology*, **48**, 87–96.

Holmes, J. C. (1962b). Effects of concurrent infections on *Hymenolepis diminuta* (Cestoda) and *Moniliformis dubius* (Acathocephala). III. Effects on hamsters. *Journal of Parasitology*, **48**, 97–100.

Janovy, J. J., Jr. (2002). Concurrent infections and the community ecology of helminth parasites. *Journal of Parasitology*, **88**, 440–5.

Wardle, R. A. and McCleod, J. A. (1952). *The Zoology of Tapeworms*. Minneapolis, Minnesota: University of Minnesota Press.

Selected reading

Anderson, R. C. (1980). *Fifty Years of Helminthology*. Fiftieth Anniversary Commonwealth Agricultural Bureau Special Volume. Toronto, Canada: Commonwealth Agricultural Bureau.

Anderson, R. C. (1992). *Nematode Parasites of Vertebrates. Their Development and Transmission*. Cambridge, UK: CAB International.

Kabata, Z. and Arai, H. P. (1997). In memoriam. Leo Margolis. (1927–1997). *Journal of Parasitology*, **83**, 619.

Roy's Students. (2002). In Memoriam. Roy Clayton Anderson. (1926–2001). *Journal of Parasitology*, **88**, 829–30.

7

The importance of field parasitology in the global assault on human parasitic disease

Diseases crucify the soul of man, attenuate our bodies, dry them, wither
them, shrivel them like old apples . . .

ROBERT BURTON, *ANATOMY OF MELANCHOLY* (1621–50)

Epidemiology is the study of disease in humans. To deal effectively with
the epidemiology of a parasitic organism and the disease it may cause,
one must first have a full understanding of the parasite's life cycle. Once
this is accomplished, then an epidemiological investigation can take any
of several approaches. These can range from mathematical modeling and
statistical analyses to those that employ modern molecular and genetic
technologies. By whatever means, however, the overall goal is to understand the ecology of the disease process, and this must always include,
without regard to the parasite species, work in the field.

The earliest record of descriptive epidemiology can be found in
Egyptian, Indian, and Chinese writings from more than 2500 years BC.
The Greeks, e.g., Hippocrates, and early Romans, e.g., Pliny, were familiar with tapeworms, the anemia pallor associated with hookworm, and the
fevers produced by malaria. Most scholars would now agree that malaria
played a significant role in the downfall of the Roman Empire. The import
of slaves from Africa to Rome unquestionably meant the introduction of
Plasmodium spp. into that part of the world. I read something not long ago
that recent DNA studies on the skeletal remains at ancient Roman burial
sites strongly suggest the presence of falciparum malaria. Arabs, in the latter part of the first millennium, AD, knew and wrote about a number of
diseases that were unquestionably caused by parasitic organisms. Descriptions of both quartan and tertian fevers, usually referred to as ague, appear in the writings of Dante (1265–1321) (*The Inferno*), Chaucer (1342–1400)

(*Canterbury Tales*), and Shakespeare (1564–1616) (*The Tempest*). According to F. E. G. Cox (writing in Topley and Wilson's *Microbiology and Microbial Infections*), at least 28 species of parasitic helminths had been firmly documented in humans by the start of the twentieth century. In 1999, David Crompton, in the *Journal of Parasitology*, indicated that 342 species of parasitic helminths had been reported in humans. Even though the great majority of these parasites must be considered as purely coincidental acquisitions, Crompton's number does not include the dozens of species of parasitic protozoans that have also been described from humans. We can justifiably conclude from these observations that humans are highly vulnerable to infection by a wide variety of eukaryotic organisms.

Over a period of about 75 years in the last half of the nineteenth and the early twentieth centuries, the life cycles of most of the major parasites that produce serious disease in humans became known. Perhaps the most dangerous from the perspective of morbidity, mortality, and economic cost, were (and still are!) *Plasmodium* spp., the causative agents of malaria. In 1880, Charles Laveran was the first to make a connection between *Plasmodium* parasites in the blood of humans and malaria. Sir Patrick Manson, generally regarded as the father of tropical medicine, urged Ronald Ross to focus on the mosquito as the vector for *Plasmodium* spp. Ross was successful in this effort (1897) and won his Nobel Prize in physiology in 1902 for this awesome discovery. Despite the significance of Ross' finding, however, it required another 50 years before Henry Shortt and P. C. C. Garnham in England conclusively demonstrated that a liver phase was necessary prior to the blood phase in the vertebrate host during the life cycle of the malarial parasite. The time gap in successfully elucidating the full cycle was due, at least in part, to a "gaff" propagated by Fritz Schaudinn, who had reported in 1903, incorrectly, that the sporozoites of *Plasmodium* spp. penetrate red blood cells directly. Ross' identification of the vector for *Plasmodium* spp. was partly stimulated by Manson's discovery in 1877 that mosquitoes were vectors of *Wuchereria bancrofti*. In the last decade of the nineteenth century, David Bruce provided substantive information on the relationship between nagana in domestic ungulates, *Trypanosoma brucei*, and tsetse flies; making another significant breakthrough in vector biology and parasite transmission. von Siebold had determined the life cycle of *Echinococcus granulosus* as early as 1853. The cycle of *Taenia solium* was worked out by Kuchenmeister in 1855, for *Trichinella spiralis* by Virchow and Leuckart in 1858, *Ascaris lumbricoides* by Grassi in 1881, *Ancylostoma duodenale* by Looss in 1898, *Trypanosoma cruzi* by

Chagas in 1909, *Schistosoma japonicum* by Miyairi and Suzuki in 1915, and *Diphyllobothrium latum* by Janicki and Rosen in 1917.

Following each of these major discoveries came the real opportunity to investigate properly the epidemiology of the parasites and the diseases they cause. Many of these findings were to have a significant impact on the control of the parasitic diseases, even to the point of developing important new drug therapies. Notwithstanding these huge successes, the prevalences of a number of these parasites are again on the rise in several parts of the world. These increases can be attributed to many factors, but in most cases the reasons are as old as the parasites and the diseases they induce. Thus, for example, we know the tropics and subtropical areas of the world possess environmental conditions that are conducive to the enormous diversity of parasitic organisms, and to their transmission. However, there are other explanations too. In addition to the simple fact that many Third World countries are in tropical and subtropical parts of the world, poverty in most of these countries continues unabated. Substandard socioeconomic conditions have always been a partner to parasitic disease. Exacerbating the poverty problems in many countries are political and social unrest, often leading to serious conflict and the mass movement of people, both of which contribute to the increased transmission of human parasites. A recent (2002) press release from the World Health Organization (WHO) states that "mass population movements are fueling the growing epidemic of 'black fever,' better known as kala azar, or visceral leishmaniasis," in a number of countries in the developing world since 1993. According to the same report, "an outbreak [of kala azar] in Sudan killed 100 000 in an area with a population of less than one million," a hefty level of mortality. Poor agricultural practices are also expedient for a number of these parasites. Finally, the development of drug resistance has become a real anathema in the treatment of some parasites, and this too has a serious impact on the spread of many diseases caused by protozoans and helminths not only in humans, but also in many domesticated animals holding great economic importance for humankind.

The nuances of successful epidemiological research are sometimes very subtle. I refer you to Chapter 4, where I detailed Will Cort's discovery of the cause of swimmer's itch in 1926. Cort had actually described the cercaria of *Schistosomatium douthitti*, the most common agent of swimmer's itch, in his doctoral dissertation published in the *Illinois Biological Monographs* way back in 1915, but he was totally unaware at the time of this parasite's relationship to swimmer's itch. Moreover, there is no question that he had acquired the disease several times prior to his elucidation of its etiology in 1926. Yet, it took him nearly a dozen years to make a connection between these cercariae and swimmer's itch. A little serendipity was even involved in making the discovery. The balance of the summer in 1927 when he made the discovery, and several summers that followed, were then spent in developing a more thorough understanding of the various epidemiological parameters of swimmer's itch.

Modern epidemiology crosses many disciplines, some biological and some not. For example, one of the most powerful tools presently employed relies on sophisticated mathematical modeling. Among the first to successfully use a modeling approach was Nelson Hairston in a study of *S. japonicum* in the Philippines in the early 1960s. Much of the modeling now involved in studies on the epidemiology of eukaryotic parasites, especially the helminths, can be traced to the two quantitative papers of Harry Crofton published back in 1971. It is interesting that Crofton's model, and the epidemiological applications that developed subsequently, were based on a very simple study of a duck acanthocephalan, *Polymorphus minutus*, in its amphipod intermediate host, *Gammarus pulex*, by H. B. N. Hynes and W. L. Nicholas in 1963. It remained for Roy M. Anderson, Bob May, Andy Dobson, and several others to take mathematical modeling to a substantially higher level in the 1970s and 1980s. Most of these modeling efforts centered on Crofton's ideas regarding dispersion. A reminder: it was Crofton's view that parasites, certainly most adult helminths, are contagiously distributed within a host population. The idea here is that a preponderance of the parasites within a given population will be found in a relatively small number of hosts. This means that only a few hosts in a population will be radically impacted by the pathogen.

Deriving from this concept are several critical questions related to the manner in which these distributions arise, especially among the geohelminths. These parasites include those with direct soil transmission for humans, i.e., *Necator americanus*, *Ancylostoma duodenale*, *Ascaris lumbricoides*, and *Trichuris trichiura*. For example, is there predisposition to infection among certain individuals within a population? If there is predisposition, then are the genetics of the parasites or their hosts linked to it in some fashion? Does host behavior, learned or unlearned, predispose certain individuals to infection, and so on? The question of predisposition is an important one, for a number of reasons (see Holland and Kennedy, 2002). Sarah Williams-Blangero and John Blangero, in a review in this book, state that several epidemiological investigations have shown that susceptibility to these three geohelminths is usually aggregated within certain families. They continue, "This evidence, in combination with the many empirical observations that worm burden is overdispersed . . . and the fact that certain individuals have a tendency to repeatedly develop high worm burdens after anthelminthic therapy, suggests that genetic factors may play an important role in determining the risk for helminthic infections."

Whereas overdispersion is a fact in virtually all helminth studies published to date, genetic predisposition is not. It has not been demonstrated in all cases. For example, predisposition (of any sort) was not shown in the 1981 Croll and Ghadirian study in Iran, or in the 1995 Kightlinger *et al.* investigation in several Madagascar villages. In both of these studies, and several others not cited here, the worm intensities were determined in a given village, then the residents were dewormed. Following an appropriate time interval, investigators returned to the villages and re-examined the residents for worm intensities. Genetic predisposition would predict that the same villagers with high parasite numbers the first time should be heavily infected the second time, but this was not the case. Thus, in both the Iranian and Madagascar investigations, new individuals carried the heavy infections. In contrast, Williams-Blangero and Blangero summarized in the Holland/Kennedy book an exceedingly thorough and exceptionally well-designed investigation they conducted in Jiri, a rural part of eastern Nepal. Without going into any details of the design, their results reveal that, "The evidence consistently indicates that there are significant genetic influences on susceptibility to *Ascaris, Trichuris, Necator americanus,* and *Ancylostoma duodenale* infections in humans. It is likely that at least 30% of the total variation in worm burden measures observed in human populations is due to innate genetic factors relating to resistance." Why the discrepancy between the Nepal results and those in Iran and Madagascar?

It is difficult to know at this point why data generated in field studies are so variable. However, perhaps part of the answer rests with another issue raised in the Williams-Blangero/Blangero review. They noted that "the potential for interaction between genomes of the human host and the helminthic parasite are enormous." The authors point out, for example, that parasites have evolved ingenious ways of avoiding the host's immune capabilities. Some parasites are sequestered and hide from the host's immune system, others suppress it, and still others incorporate host antigens into their own surfaces as a way of mimicking the host. Williams-Blangero and Blangero hypothesize "that genetic variation present in the parasite interacts with genetic variation present in the human host to *jointly* [my emphasis] determine the observed variation in quantitative measures of helminthic worm burden," – a really neat idea. But, as they note, the hypothesis "remains to be tested." This conceptualization rests squarely on a co-evolutionary platform. For me at least, it will be of more than just a passing interest to see if protocols can be designed and executed to test this notion.

Modeling efforts have given rise to several new ideas regarding the most efficacious way of treating some of these diseases. For example, epidemiologists and others began to question the necessity of using mass treatment for a number of parasitic helminths. The models suggested that there might be some abundance threshold below which parasite transmission dynamics can be disrupted. If the parasite's numbers in a host population can be reduced below this threshold, whatever that might be, then predictions indicate the parasite should disappear locally. Based on these hypotheses, proposals have been made that drug therapy would be effective for an entire village if only the most heavily infected individuals receive therapy (selective treatment), or, alternatively, if just those with behaviors conducive to parasite transmission receive what is referred to as targeted treatment.

Whereas drugs for treating most helminth parasites are effective and readily available, they can be relatively expensive, certainly for residents of Third World countries. Moreover, health care delivery is a serious problem, especially in rural areas of countries where many of these helminth parasites are most commonly encountered. Consequently, mass treatment is both difficult and costly. Treating just those with the heaviest parasite loads (selective treatment) is also expensive because the heavily infected individuals must first be identified, requiring stools to be collected and eggs counted. This is labor-intensive work and adds significantly to the costs. Targeted treatment requires a precise knowledge of the parasite's epidemiological characteristics, which may vary locally. Ideally, the most effective way of handling the problem of health care would be to raise living standards in Third World countries, to provide better sanitation, to use better agricultural practices, and to educate local populaces as to the nature of their endemic diseases. For the most part, these outcomes are not likely soon, and probably not even in the distant future. The reality is that prospects for reducing the transmission of most of these helminth parasites are dim at the present time, despite the fact that recent WHO reports indicate there have been a few successes, e.g., dracunculiasis

(*Dracunculus medinensis*) and river blindness (*Onchocerca volvulus*). An even more recent WHO press release, however, indicates that there was a recent increase in dracunculiasis in the Sudan, probably in conjunction with the 20-year war that persists to this day. So, even with this disease, we are not yet at the point of eradication or of declaring success.

Another important consideration in the elimination/reduction of parasitic disease is based squarely on national priorities. Consider malaria as an example. Philip Rosenthal and Louis Miller, writing in Rosenthal's recent book, *Antimalarial Chemotherapy,* state that, "As a disease of the poorest nations, malaria remains a poor cousin of the major health problems of the developed world, and funding for malaria is dwarfed by that for heart disease, cancer, AIDS, and asthma." Estimates on mortality due to malaria vary, but are generally placed at between two and three million annually. In sub-Saharan Africa alone, it is believed that approximately 1 million children under the age of 6 die each year, mostly from falciparum, or cerebral, malaria. Despite these facts, Rosenthal and Miller cite data which indicate that the US expenditure in 1990–92 in dollars per fatality caused by AIDS was $3274, for asthma it was $789, and for malaria, a paltry $65. Why the disparity? The answer is fairly clear. AIDS is devastating in Africa, with an estimated 2.3 million dying just in 2001 alone. But, it is also relatively prevalent in all of the developed countries of the world, including the USA, though the extent of mortality is not even close to that occurring in Africa. In contrast, malaria is also devastating in Africa, but it rarely occurs now in any of the developed countries. Health care priorities are based on the perceived national needs in a given country. In developed countries, AIDS, and even asthma, are more significant problems than malaria, so it is easy to see where and why the dollars will be spent. We all know about squeaky wheels. Those who make the most noise or carry the greatest political clout will be the ones heard by the so-called decision makers in the government in the USA and elsewhere.

When the USA was involved in World War II, relatively large sums of money were spent on malaria research because many of our military personnel were in malarious parts of the world. After that war was over, the research dollars for malaria declined. The same thing was true during the Korean conflict and in Vietnam because, in both places, malaria was a problem and thus became a national health care priority for the USA. I recently finished reading the really excellent book, *We Were Soldiers Once . . . And Young* (1992), by Lt. General Harold Moore (Ret.) and Joe Galloway. It is an account of a fierce battle between the First

Battalion, Seventh Cavalry Regiment of the First Cavalry Division and North Vietnamese regulars in the Ia Drang Valley in South Vietnam in 1965. One of the things that struck me was the large number of US soldiers who were affected by malaria before the battle even took place. Galloway, a print reporter who witnessed the fight, and General Moore (then a Lt. Colonel and commander of the First Battalion) said in the book, "Once each day every man of us, under close supervision, choked down a bitter, dime-size yellow malaria pill. It was an automatic Article 15 offense – one subject to nonjudicial punishment – to be caught sleeping outside one's mosquito net, no matter how hot it was. Even so, we began losing men to malaria within two or three weeks. Within six weeks fifty-six troopers from my battalion alone had to be evacuated to hospitals, suffering serious cases of malaria. The problem was a particularly virulent falciparum strain of the disease, prevalent in the Central Highlands; it was resistant to the antimalarial drugs available to us at the time" (and it probably still is!).

The loss of 56 men represented approximately 10% of the battalion's fighting strength, all within about 2 months of arriving in South Vietnam. These numbers were typical in that conflict. They reflect the huge impact of an unseen, yet always present, parasitic enemy. According to the Moore–Galloway book, however, the North Vietnamese soldier was equally affected by the undiscriminating *Plasmodium* parasite. Moore and Galloway described how the North Vietnamese infiltrated from the north in battalion-size groups, walking all the way down to the south. In addition to their food and military equipment, all of them carried 50 antimalarial pills, enough for one each day they were on the trail. Before reaching their destination, every single man would be suffering the effects of this dreadful disease and, "on average, three to four soldiers of each 160-man company" perished during their long trek because of malaria.

For the US combat infantryman and marine, this terrible scourge has had a major impact on virtually every military campaign from the Revolutionary War to Vietnam. During the Civil War alone, some authorities have estimated that malaria caused more hospitalizations than rifle bullets or cannon shells, and that was the bloodiest war in which Americans have ever participated. If one thinks in terms of the enormous investment required to develop the smart-bomb technologies, stealth bombers, and other sophisticated weaponry, one must also wonder why the US government has not simultaneously placed greater emphasis on dealing with this parasite in a more effective manner. Dodging bullets is tough enough, but it is impossible to avoid nagging mosquitoes carrying the dread malarial parasite!

I do not wish to minimize the importance of AIDS, asthma, or tuberculosis research. These are all horrible diseases and, therefore, of major public health concern. I have attempted to make the case that neither malaria nor most of the other tropical diseases affecting the Third World have very high priorities with respect to biomedical research in the developed countries. Personally, I think this is one of the major tragedies of our time and I am confident this feeling is completely shared by anyone with a modicum of knowledge regarding the problem. History always provides major lessons from which to learn. On occasion, we do. In the case of malaria, the lesson we have learned is worth about $65 per fatality. This is shameful!

Teams of scientists have recently managed to sequence the genomes of *P. falciparum* and *Anopheles gambiae*, the primary vector of this parasite in Africa. The objective in these efforts is clear, i.e., to open the door even wider to the development of vaccines and new drugs for cerebral malaria, or the creation of transgenic mosquitoes that are unable to transmit the malarial parasite. However, as Declan Butler wrote in *Nature* in 2002, "the availability of the *P. falciparum* genome does not herald the parasite's impending doom. The biggest bottlenecks often lie downstream, in getting funds and industrial expertise to move promising drugs and vaccines from the lab to the market." He continued, "With the resources currently available, say health economists, talking about defeating malaria is like promising to build a $100-million skyscraper with just $1 million in the bank. The stark truth is: if we don't bankroll the effort, we won't roll back malaria." It is the same old sad song. In summary, the bottom line is the absence of adequate financial commitment by the major industrial countries and pharmaceutical companies of the world, and it does not look like major change is in the forecast either.

The worldwide impact of diseases produced by eukaryotic organisms is significant, whether measured in terms of treatment cost, mortality rates, or morbidity. In 1993, the World Bank published the results of an impressive study in which they attempted to quantify the effect of these diseases in different parts of the world. They transposed these negative impacts into units, called disability-adjusted life years (DALYs). In Third World countries, DALYs due to infection by eukaryotic parasites represented 37.7% of the total losses; in developed countries, the total was nearly an order of magnitude smaller, about 4.6%. Whereas these parasite problems are primarily associated with Third World countries, they also vary substantially from one part of the world to another. In Africa, despite the geohelminths, trypanosomiasis, and leishmaniasis, the significant disease caused by eukaryotic parasites is malaria. In China and several other Asian countries, the difficulties rest mostly with enteric helminths, e.g., hookworm disease, ascariasis, and trichuriasis. In Latin America, Chagas' disease is the prime problem, although the complications caused by enteric helminths are a close second. Throughout Third World countries, these diseases also vary with age. In the youngest age group, i.e., <5 years, malaria is a prime cause of health problems. In the age group from 5 to 14, the difficulties are mostly associated with enteric worms.

Cost is not always the central issue in dealing with some diseases. As I mentioned previously, there are drugs, e.g., praziquantel, mebendazole, albendazole, and ivermectin, which are highly effective for most helminth parasites. For many of the protozoan parasites, drugs also are available, but resistance has become a serious problem. For some of the other protozoans, *T. cruzi* for example, there is no really effective treatment. M. A. Miles (in Volume V of Topley and Wilson's *Microbiology and Microbial Infections*) states that, "Cure rates with both nifurtimox and benznidazole (for Chagas' disease) are not total and vary regionally." Continuing, he says, "The chronic phase is seldom treated...." Treatment for several other parasites presents different sorts of special difficulties. Consider hookworms, for example. It is well known that adult hookworms are intestinal and are vulnerable to mebendazole. Whereas therapy applied to the adult worms may be effective, it does not work against those L3s in a state of developmental arrest that may be sequestered in the tissues of human hosts. Gerhard Schad, Peter Hotez, and others have pointed out that these tissue-dwelling larvae will, at some point, become activated and continue their migration into the intestine where they become adults. The problem here is simple: one may treat the adults and cause them to be eliminated, but as soon as the sequestered L3s move, the host will acquire adult worms again. This is a wonderful example of what I call an "epidemiological ruse." You think you have the parasite under control, but it has a clever way of evading what should be the knock-out punch.

A number of enteric helminths in domesticated animals, i.e., sheep and goats, have developed resistance to anthelmintics such as the benzimidazoles. A review by Derek Wakelin and Janette Bradley, written for the Holland/Kennedy geohelminth book, details the problem. They cite a 1996 paper by W. N. Grant and L. J. Moscord, which refers to resistance of *Haemonchus contortus* and *Trichostrongylus colubriformis* to the benzimid-azoles because of a point mutation which results in the placement of tyro-sine rather than phenylalanine at position 200 in ß-tubulin isotype 1 gene. Wakelin and Bradley go on to note that "a significant aspect" of this resis-tance "is the speed with which the mutant gene, once selected, has spread through the nematode populations." This is fascinating, but scary, stuff because it has wide-ranging ramifications in developing adequate drug protocols for a critical industry in many agricultural areas of the world.

Treatment of a disease caused by a parasite in which a human is the only host may be difficult, but it is not impossible or impracticable if effective drugs are available. Treatment of a zoonotic parasite, in contrast, creates greater problems in many cases. For example, a few of the simian malarias are known to be transmitted to humans. Can the monkeys in a jungle be treated? Should the monkeys in the jungle be eliminated? Both are absurd questions. In one of the first modeling efforts back in the 1960s, Nelson Hairston estimated that if *S. japonicum* was eradicated in a rural Philippine village, within a year it would return to pretreatment levels. Why? Be-cause rice rats are reservoir hosts for the parasite and rice rats are every-where in the rural Philippines. The same questions can be asked again. Can, or should, rice rats be treated or eliminated? The answer is the same as for simian malaria; both are absurd questions. *Cryptosporidium parvum* uses cattle as a reservoir, *Toxoplasma gondii* is normally a parasite of felines, *Giardia lamblia* and the several species of *Trichinella* occur naturally in a range of mammals, and I could go on. Of course, all of these parasites can be easily transmitted to humans. The issue is simple. Zoonotic diseases are in many ways almost intractable because, even if they are treated success-fully and eliminated from the human host, the parasite always lurks in a reservoir and can be easily recruited again, and again.

One of the more useful approaches to understanding the epidemiol-ogy of parasitic disease has been with the development of molecular and genetic techniques in recent years. The earliest application in the epi-demiological arena involved the use of relatively simple starch gel elec-trophoresis to compare the differential mobility of enzymes and other proteins in various species. Linus Pauling and Harvey Itano employed this technique for elucidating the differences between normal and sickle cell hemoglobin as early as 1949. Newer procedures have much greater preci-sion for measuring genetic variability. They carry acronyms such as RFLP, RAPD, SSCP, CFLP, and others. These technologies are generally used to compare nucleotide sequences after restriction enzymes have cut the DNA molecules into small pieces. Hybridization and amplification (polymerase chain reaction) techniques may be applied to copy or synthesize specific DNA fragments, or entire nucleotide sequences of DNA.

A multitude of parasitic organisms, ranging from leishmanial parasites to *Echinococcus* spp., have been examined using these genetic/molecular procedures. Let's consider the latter organisms as an example. Presently, most workers recognize four species of *Echinococcus*. These include *E. granulosus, E. multilocularis, E. vogeli,* and *E. oligarthrus.* In the volume, Echinococcus *and Hydatid Disease,* edited by R. C. A. Thompson and A. J. Lymbery (1995), D. P. McManus and C. Bryant provide a cogent review regarding the biochemistry, physiology, and molecular biology of these four species. In another chapter, A. J. Lymbery discusses, just as clearly, the genetic diversity, genetic differentiation, and speciation of parasites in this genus. According to McManus and Bryant, eight strains of *E. granulosus* are recognized based on epidemiological and biological considerations, whereas Lymbery lists seven genotypes as determined by mitochondrial DNA sequencing. One of the interesting observations by these authors is that all of the strains employ canines as their definitive hosts, yet the larval stages are highly adapted to specific mammalian intermediate hosts. Thus, in one part of the world, hydatid cysts are found primarily in cervids, in sheep in another, camels elsewhere, and so forth. In the UK, there are distinct horse–dog and sheep–dog strains, which vary widely in a range of biological characters. Within each of these strains there appears to be a great deal of genetic uniformity, but considerable genetic differences between them as well. According to J. Bowles and D. P. McManus, however, there is no evidence for gene exchange between the strains, despite occurring sympatrically and using the same canines as definitive hosts.

These genetic differences in *E. granulosus* were brought home to me in a personal way, long before application of the newer molecular genetic techniques to parasites. In the academic year of 1971–72, I took a leave from Wake Forest University to work in the laboratory of the late Desmond Smyth at the Imperial College of Science and Technology in London. My reason for going there was to learn something about the in vitro culture of cestodes. I could not have picked a better individual than Desmond Smyth with whom to work. Not only was he the world leader in this field, he was also an absolutely marvelous person. He had just moved to Imperial College from Australia where he had gone several years previously to set up a new Zoology Department at the Australian National University in Canberra. It was while he was in Australia that he developed a successful in vitro culture procedure for *E. granulosus*. My friend, Ray Kuhn, and I had been working with *Taenia crassiceps* for a several years and I wanted to try my hand at culturing this cestode in vitro, so I went to work with Smyth.

The first few weeks in London were spent learning sterile technique, how to prepare the biphasic culture vessels, and other in vitro culture procedures. Smyth was very patient with me as I stumbled through the process. Then, after I had learned what he thought I needed to know, he announced we were ready to start. I was either exceedingly lucky, or his system was exceedingly versatile, or both, because the first time we used it for *T. crassiceps*, it worked beautifully. In fact, the technique worked just as well for *T. crassiceps* as it had for *E. granulosus*, and we continued to have success almost every time we tried it. The only problems I had would occur when, on occasion, I clumsily contaminated a culture vessel while changing the liquid phase of the medium.

During the first few months of my stay, Smyth was busy getting settled into his new environment. In addition to giving me in vitro culture lessons, he worked hard on editing the (then) new *International Journal for Parasitology*, of which he was the Founding Editor. He was also occupied in teaching his course in parasitology. One day, he announced that he was ready to do some bench work. So, he headed for a local abattoir where he was able to obtain horse livers with hydatid cysts. It was the first time that I had ever seen these parasites in situ and I was quite taken by the huge numbers of brood capsules and the exquisite protoscolices inside. Most of the hydatid cysts were about the size of a grapefruit and were usually sequestered just below the surfaces of the livers. He and Zena Davies, his technician, set up the culture media and flasks, and followed exactly the same process that he had developed successfully in Australia for *E. granulosus*, the same one that I had been using to grow *T. crassiceps*. But, his initial effort was a failure. The protoscolices just sat in the culture flasks and did nothing in terms of development. He and Zena tried his technique over and over, and they still could not make it work, despite the fact that I was continuing to grow *T. crassiceps* at will using the same system.

After a number of unsuccessful efforts, one could see his frustration begin to set in. At this juncture, he was about to affirm my earlier belief that in vitro culture of helminth parasites was an unequal mixture of science and alchemy, mostly the latter! Casting about for an explanation, Smyth variously considered if there was a problem with the local water, the serum source, the culture medium itself, and several other potential pitfalls. He pursued just about every conceivable aspect of his culture system, but nothing was successful. Then, it dawned on him. The hydatid cysts he used in Australia were from sheep and those in London were from horses. He wondered, were they different? So, he obtained hydatid cysts from sheep in the UK and introduced them into his culture system. They grew just like the sheep *E. granulosus* he had used in Australia to develop the culture system for the parasite. Without really trying, they had generated indisputable, first-time evidence for a basic physiological difference between hydatid cysts in horses and those in sheep – a finding that, without question, reflected the genetic differences between strains that were later to be confirmed using molecular techniques.

Smyth and Zena Davies published their observations in the *International Journal for Parasitology* in 1974. This failure/success was responsible for ultimately opening the door to understanding strain variation in the epidemiology of hydatid disease. As noted by R. C. Andrew Thompson and Donald McManus in a *Trends in Parasitology* paper in 2002, "Desmond Smyth demonstrated fundamental differences in the developmental biology of *E. granulosus* of sheep and horse origin. These seminal studies provided a platform of understanding for both earlier and subsequent observations on differences in development and infectivity between isolates of the parasite from various host species in different parts of the world." I have often thought about how much fun it was in being present in Desmond's laboratory and in seeing first-hand the highly significant breakthrough in this area of parasitology.

My personal attraction to the large bladder worms of taeniid cestodes began when I undertook my very first research as an undergraduate student at Colorado College in the fall of 1957. It involved the larval stage of the cestode, *Taenia multiceps*. The life cycle of this tapeworm includes the coyote, *Canis latrans*, as the definitive host, and the black-tailed jackrabbit, *Lepus californicus*, as its intermediate host. Adult cestodes shed proglottids in which eggs are found. Once outside the host, eggs escape from the gravid proglottid – rather easily I was later to discover. Jackrabbits accidentally ingest the eggs, which hatch in the small intestine. A hexacanth embryo emerges, penetrates the gut wall, and migrates to a subcutaneous or intramuscular site via the circulatory system, from which it exits and develops into a very large bladderworm, called a coenurus. When a coyote kills and consumes an infected jackrabbit, it becomes infected with the parasite.

When I undertook my dissertation research at the University of Oklahoma, I needed both adult and larval stages of the cestode for the physiological and biochemical work I was doing. Obtaining gravid proglottids was simple after I exposed a dog to a coenurus taken from a jackrabbit shot near Harper, Kansas, on a fine spring day in 1961. Proglottids shed were easily collected whenever the dog defecated. However, when I exposed many, many lab rabbits to eggs of *T. multiceps*, an infection was never obtained – a very distracting situation since the literature referred to cases of experimental infections in lab rabbits in other parts of the world. But, I could not make it work.

Subsequently, I was to do a post-doc in the School of Public Health at the University of North Carolina (UNC)-Chapel Hill. I took the infected dog with me to my new setting, anxious to do some more work with the parasite. A graduate student, Bruce Lang, suggested that I use eggs of *T. multiceps* to infect mice of a Swiss strain that was maintained in the Department of Parasitology where I was working. I was most dubious of this idea since mice had never been reported as hosts for *T. multiceps*. However, I decided that I had nothing to lose and went ahead with the egg intubations into a number of these animals. A few months later, I noticed what appeared to be subcutaneous tumors in several of the mice which had been fed eggs, but I did not really think much about it as this particular strain of mouse was prone to develop these sorts of tumors. I recall going to the American Society of Parasitologists meetings in Chicago that year and deciding that I would remove the tumorous mice on my return to Chapel Hill. When I killed my first mouse and skinned it, I was absolutely stunned to find a coenurus of *T. multiceps*. I fed this coenurus to an uninfected dog and, about 6 weeks later, it was shedding gravid proglottids. Eggs from these gravid proglottids were then fed to mice and coenuri developed as they had previously. I was, therefore, able to satisfy Koch's postulates, much in the same way that Roy Anderson had done with his duck filarial worm (see Chapter 6).

Something else occurred during this time though; something that was to frighten me greatly. In a number of the mice to which I had fed eggs, there were significant behavioral changes. Several had begun to run in circles, with their heads tilted to one side, and they lost weight rapidly. I killed these animals and necropsied them. Much to my surprise, and then dismay, I discovered tiny coenuri growing in their brains. My concern about the situation developed from the fact that *T. multiceps* was also the cause of gid, or staggers, in sheep, due to the presence of coenuri in their brains. To make matters worse, there were anecdotal reports in the literature that *T. multiceps* could infect humans and other primates as well, and in the brain. I surmised that if the parasite would go into the brains of laboratory mice, or sheep, or other primates, it could just as easily go into me! Moreover, I had been taking absolutely no precautions to avoid such an infection over the 5 years that I had been working with gravid proglottids and eggs of the parasite. And, to make matters even worse, about that time I was contacted by the Centers for Disease Control in Atlanta regarding a young lad in Wyoming who had a coenurus in his brain. They wanted me to confirm their diagnosis of coenuriasis, which I did and, as my wife Ann says, I "did it fretfully."

"Intern's syndrome" is a well-known phenomenon that occurs among young physicians. They sometimes acquire the same signs and symptoms suffered by patients they are treating. Though not a physician, I was to develop a serious case of intern's syndrome. Every time I had a headache, I wondered about its cause. I can even recall going home at night on occasion and standing, first on one leg and then on the other, with my eyes closed and my arms extended laterally, just to see if I could do it. I had read someplace that this was a good way to test one's balance. Ann and I often enjoy a good laugh about this episode, but at the time, it was not at all funny.

Of course I did not acquire cerebral coenuriasis, but I did become interested in parasites that occur in the brain as a result. Among these is one that causes neurocysticercosis, a disease produced by the larval form of *Taenia solium*, the so-called pork tapeworm. The definitive host for this cestode is a human, in which it will grow to the length of about 15–20 foot in the small intestine. Pigs accidentally ingest eggs in the gravid proglottids shed with the feces. On hatching, a hexacanth embryo then penetrates the gut wall, is picked up, presumably, in the circulation and is carried to the final site of infection, typically the heart or somatic muscle (Fig. 45), where it develops into a cysticercus, a relatively small bladderworm approximately 10–15 mm in length. The cysticercus posseses an invaginated scolex and a small bladder filled with a transudate from host tissue fluid. When a human consumes the uncooked, or undercooked, flesh of an infected pig, the life cycle is completed. However, humans may also ingest eggs of *T. solium*. When this occurs, the cysticercus can also develop in humans, not only in the same somatic and heart muscles as in the pig, but in the brain as well (Fig. 46). If the parasite occurs in the brain, neurocysticercosis is the outcome, and severe neurological problems will develop.

Figure 45 Examination of a pig's tongue for cysticerci of *Taenia solium* in Mexico. The nodules created by the cysticerci in this piece of somatic muscle are easily felt and can usually be seen. This procedure is routinely employed to assess the presence of the parasite. (Courtesy of Mirna Huerta.)

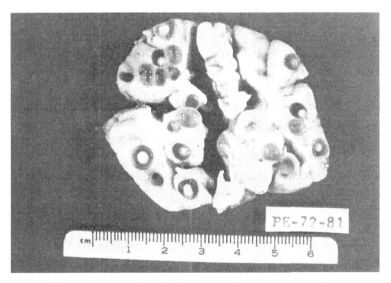

Figure 46 *Taenia solium* cysticerci in the brain of a human at autopsy. Humans accidentally acquire these larvae when they ingest the eggs of *T. solium*. (Courtesy of Mirna Huerta.)

It is, for example, the leading cause of epileptic seizures in Latin American countries of the western hemisphere. Darwin Murrell says that it is now the leading cause of epileptic seizures in India. His work with a group of neurologists in Vellore, India, has produced data suggesting that it is probably the leading agent for epilepsy on the entire Indian subcontinent, and that it has made huge inroads in creating the same level of problems in East Africa as well.

Perhaps one of the most fascinating accounts of neurocysticercosis and some of its epidemiological characteristics can be found in a wonderful little book called *New Guinea Tapeworms and Jewish Grandmothers*, written by Robert Desowitz (1981). This man is a marvelously clever purveyor of parasitological tales and has written several extremely fascinating books describing his experiences with some hugely exotic parasites and places around the world, e.g., *Malaria Capers* and *Who Gave Pinta to the Santa Maria?* In the New Guinea tapeworm book, he describes his trip into the central highlands of Irian Jaya of Indonesian New Guinea. He was sent by the WHO to investigate an outbreak of what was believed to be neurocysticercosis among the Enarotali Ekari, a group of primitives possessing, as Desowitz put it, "Stone Age toilet habits" and, among other things, some rather bad eating habits as well.

His trip to New Guinea was occasioned by reports from two physicians of a sudden rash of burn victims, perhaps 25–30 per month, at a small Enarotali hospital. The people were so badly hurt in some cases as to require amputation of their burned limbs. A fire in the center of their huts is necessary because of the extreme cold temperatures at the high altitudes of the central highlands. When questioned by local authorities, all of the patients indicated that, while sleeping, they "had fallen unconscious into the household fire." Because of the sudden increase in burn victims requiring treatment, and the similarity in the seizures, it was suspected that "some new infectious agent had been introduced." Seeking an explanation for the new and dangerous problem, scientists in the Department of Parasitology at the University of Indonesia in Djakarta examined stool samples from the local villagers. They were surprised to find that 8% of the samples had eggs of a taeniid cestode, the first time a parasite such as this had been seen in New Guinea. A more careful physical examination of the Ekari people revealed the presence of subcutaneous cysticerci. The identity of the parasite was confirmed as *T. solium* on finding cysticerci of the parasite in the brain of a person who had died. Subsequent serological investigation indicated that at least 25% of the Ekari adults and children had cysticercosis. But why the sudden appearance of the disease? As described by Desowitz in his inimitable way, "Reconstruction of historical events indicates the tapeworm came unseen, riding an anticolonial wave; the vehicles of transport were men and pigs."

It seems that, in 1969, the peoples in western New Guinea were given the opportunity of joining the State of Indonesia by the United Nations. In association with a local plebiscite, Indonesia's President Suharto placed troops into Enarotali. The military occupation included soldiers from Bali who were Hindu in religion, in addition to Muslims, the predominant religion in Indonesia. Suharto also presented the Ekari with a gift, pigs from Bali. Unfortunately, the pork tapeworm is endemic to Bali, although neurocysticercosis is largely absent because the Bali natives, according to Desowitz, "are fastidiously clean in their personal habits." The appearance of the tapeworm, in concert with the extremely primitive living conditions of the Ekari, was all that was necessary to create the perfect opportunity for transmission of the parasite from pigs to humans, and from humans to humans. Desowitz went on to explain his total sense of frustration with the situation, especially in view of the knowledge that the Ekari people were fully aware that the uncooked, or poorly cooked, pork was responsible for their seizures. As a local village chief explained to Desowitz, they were required by custom to eat pork in this way. "If a child is born at night we must sacrifice a pig immediately; there is no time to look and see if it has the seeds [the parasite]. The pig must be killed and eaten at once," explained the chief.

The problem of the Ekari is classic with the native peoples in Third World countries where pigs are raised in rural areas without appropriate sanitation procedures or facilities. A few years ago, I had the opportunity of spending several days with Carlos Laralde and Edda Sciutto (a husband-and-wife team) at the Institute for Parasitology at the National University of Mexico, and Ana Flisser, who heads the Mexican equivalent of the US Centers for Disease Control. All three of these parasitologists are world-class experts on neurocysticercosis. I was met by Carlos and Sergio, the Institute's chauffeur, at the Mexico City airport, and driven to Cuernavaca where I was put up at the Las Quintas, an absolutely posh, five-star resort. In Cuernavaca that evening, I was treated to a wonderful dinner cooked by Carlos and his charming wife, Edda, in their lovely home. The next day, I took a drive with Carlos and Edda out into the countryside where we visited one of the nearby Aztec archeological sites on our way to the town of Taxco. By the way, Taxco is quite a famous place in Mexico because it is the site of silver mines used by Montezuma before the arrival of Cortez. We were there on a Sunday and the atmosphere was absolutely spectacular. Hundreds of cheerful people milled in an open square near the center of town, many of them vendors, selling their marvelous handmade wares. There was a stage and a Mexican band played happy music, to which young teenagers wearing bright native costumes danced and entertained the people who gathered around the platform. A magnificent cathedral stood on one side of the crowded square. The old church was most impressive. It was filled with beautiful ornate carvings and all sorts of exquisite statues gilded in gold. It was truly a fantastic outing. The hospitality shown me by Carlos, Edda, and Ana, and their students and other colleagues, was really awesome, something I will never forget!

But now, back to parasitology in the field. During our trip to Taxco, we passed through several of the local barrios. Pigs were running loose everywhere, or so it seemed (Fig. 47). As we drove into one of the villages, Carlos pointed to a Catholic church and explained that, on Sundays, the local populace would gather for their religious services. After leaving, there would be human fecal droppings scattered all about the church and the pigs would come scrounging for food, including the human

Figure 47 Pigs wandering free in a Mexican barrio. Humans become infected with the adult cestodes when they consume raw or partially cooked pork. (Courtesy of Mirna Huerta.)

feces. The wind kicked up a cloud of dust as we passed near the church, and I wondered aloud, "Would eggs of *T. solium* be in that cloud of dust?" Carlos smiled and replied, "What do you think?" When I teach my under-graduate parasitology course now, I tell my students this story as a way of illustrating the subtleties in the epidemiology of neurocysticercosis in particular, and of tropical diseases in general.

Based on these accounts, one would think that neurocysticercosis is strictly a disease associated with Third World countries and poverty, but this is not the case. For example, in recent years there has been a substan-tial increase in Hispanic migration into the USA, both legal and illegal, from south of the border with Mexico. Whereas there are any number of discordant social manifestations related to this mass movement, sev-eral public health problems have also appeared, especially in the south-western USA New reports indicate that nearly 25% of the hospital beds devoted to neurological problems in Los Angeles, California, are related to cysticerci of *T. solium*. According to Peter Hotez, "Neurocysticercosis is emerging as a leading (if not THE [his emphasis] leading) cause of epilepsy-seizures (other than childhood febrile seizures) among children living in Los Angeles as well as other southwestern cities."

As I was writing the essay, I originally intended to state that most of these neurocysticercosis problems in the south-western USA are the re-sult of parasite acquisition in the home countries of these Hispanic mi-grants, but I would have been at least partially incorrect. I was reminded by Peter Hotez of a situation that occurred in the early 1990s, and one of the most bizarre accounts of neurocysticercosis in the USA, or anywhere else for that matter, of which I am aware. The incident was described in a paper written by Peter Schantz and several colleagues for the *New England Journal of Medicine*. As was previously indicated, *T. solium* is strictly a par-asite of pigs and humans. In other words, there are no reservoir hosts for this tapeworm. Accordingly, one should certainly not expect to see neurocysticercosis in people of the Jewish faith because of their exacting religious ban on the consumption of pork. However, the unthinkable, but completely orthodox, transmission of *T. solium* was found to occur in, of all places, a community of orthodox Jews in New York City! According to Schantz *et al.*, between June 1990 and July 1991, *T. solium* was diagnosed by brain biopsy or immunoblot assay for antibodies in four unrelated res-idents of this Jewish enclave. None of these individuals had ever eaten pork and only one had traveled in a country where *T. solium* was endemic. Six women, all domestic employees and all recent immigrants from Latin

American countries where *T. solium* is endemic, were tested by Schantz and his colleagues for the presence of the parasite. One was infected with an adult tapeworm and another was positive for *T. solium* antibodies. The housekeeper with the adult cestode, not unlike the infamous "typhoid Annie," had unwittingly distributed eggs of the parasite to four Jewish residents in one of the most urbanized areas of the world, and certainly well above the poverty line in comparison with any Third World country. I think this particular epidemiological tale illustrates as well as any I know the sometimes strange subtleties that may be involved with the transmission of parasites to humans.

I am reminded of yet another unusual story that deals with an uncanny aspect of parasite transmission and epidemiology. This one had a much more tragic outcome than the New York neurocysticercosis situation. It is related to a severe outbreak of capillariasis on the island of Luzon in the Philippines, which occurred back in 1966. A significant player in that episode was my good friend, Darwin Murrell. Darwin was a graduate student in the Department of Parasitology in the School of Public Health, UNC-Chapel Hill, North Carolina, when I arrived there in the fall of 1963 to do a National Institutes of Health postdoctoral fellowship with John Larsh. When I was writing this particular essay, I remembered Darwin's involvement with the tragic *Capillaria* difficulty in Luzon and asked him to recall it for me, which he was kind enough to do. It is well worth sharing here, because, as Darwin said, "Working in the field in a rural area like Luzon . . . taught me volumes on the strength of linkages between culture, tradition and the opportunistic traits of parasites." This observation is not unlike the one described by Bob Desowitz as a result of his experience with the Ekari people in New Guinea.

While still working on his Ph.D. at UNC, Darwin was drafted into the US Navy, with the rank of Ensign. This was during the nasty Viet Nam era and many young men were being drafted into the military at that time. Darwin had some good luck, however, as he was assigned to the Medical Zoology Department of NAMRU2 (US Naval Medical Research Unit no. 2) in Taipei, Taiwan in November of 1966. His good fortune in assignment was greatly magnified by the presence of John Cross, who had been given the task of building a "comprehensive program to support the war effort in Viet Nam, and to provide greater coverage of infectious disease problems throughout Southeast Asia." Darwin continued, "My first real 'combat' with a serious parasitic disease occurred, fortunately, under the tutelage and support of this amazing parasitologist."

As Darwin tells the story, John Cross walked into the lab in Taipei one morning a few months after Darwin arrived and instructed him to pack a bag and head for Manila, immediately. Why the hurry? Well, it seemed the wife of a Philippine senator had persuaded a US Navy admiral that a parasitologist was needed to help deal with some sort of new parasitic disease that had suddenly appeared on the scene. So, the next day, Darwin and a young Navy epidemiologist, Roger Detels (who was to become a Professor at the University of California at Los Angeles), were on a plane heading to Manila where they were to meet with the senator's wife and several officials in the Ministry of Health. On the flight to Manila, Darwin and Roger naturally talked about what they might find. They had been alerted to the potential involvement of a new species of *Capillaria*, but Darwin was skeptical about the possibility of a parasitic helminth being the culprit. He remarked to Detels, confidently, "Worms cause chronic problems, not acute diseases on an epidemic scale." As Darwin continued, "Roger, deprived of my parasitological training, had a much more open mind!"

On arrival, they were confronted with what the Manila newspapers were calling, "the mystery disease of Pudoc." Pudoc is a small coastal village in the province of Ilocos Sur where a great many people were suffering from a cholera-like disease, which, it was being suggested, was caused by a new species of the nematode, *Capillaria*. This parasite was first observed during the autopsy of a patient who had died of "intractable diarrhea" in Ilocos Norte, a province about 20 km north of Pudoc where the new epidemic was then wreaking havoc on the locals. Maybelle Chitwood, Carmen Velaszuez, and N. G. Salazar had described the new parasite as *Capillaria philippinensis*, and had reported on it at the first meeting of the International Congress of Parasitologists in Rome in 1964.

When the Philippine health officials responded to the reports of several unusual deaths in Pudoc in 1967, they also discovered "that an even greater number were seriously ill, exhibiting prolonged diarrhea," which led to horrible emaciation (Fig. 48) and, ultimately, death. The residents of Pudoc were convinced that they had committed some sort of vile act and that they were under a curse, doomed to die a horrible death like their fellow villagers. After traveling by a four-wheel drive vehicle through the back country of dense forests and many streams, the team from NAMRU2 finally arrived at Pudoc. As Darwin described it, "The first site that grabbed our attention was a small shrine that had been erected next to the stump of a large tree that had been cut down. The residents

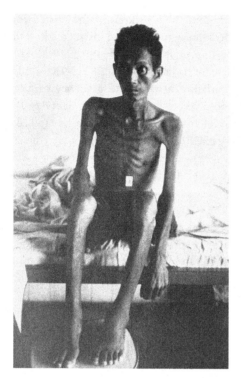

Figure 48 A Filipino man suffering the effects of "the mystery disease of Pudoc," caused by *Capillaria philippinensis*. (Courtesy of John Cross and Darwin Murrell.)

told us that after so many men and boys had died (probably beginning in 1965) they brought in a 'witch doctor' (closest English translation), who declared that the tree must have been sacred and that the barrio was being punished for cutting it down. Hence, to make amends, a shrine was set up and gifts were placed on it" (Fig. 49).

The localized epidemic had been devastating. In 1967 alone, there were 1037 cases and 77 deaths, mostly among males. So many men had died in the tiny village of Pudoc that food was in short supply and large quantities of wheat and rice, provided by the United States Agency for International Development, were being shipped in by truck. "In fact," Darwin continued, "because of the disproportionate loss of so many men (husbands and fathers), surviving men were adopting or marrying the widows and children to help them survive." He wondered, "What is going on here, still convinced that worms do not act like this."

Figure 49   The shrine built by the Pudoc people on the advice of a local "witch doctor" to atone for cutting down a sacred tree. (Courtesy of John Cross and Darwin Murrell.)

Darwin's skepticism was finally mitigated, however, by the results of clinical studies on several patients sent to a large Manila hospital. In two of those who had died, they had discovered 59 000 and 200 000 worms per liter of bowel fluid. Moreover, there were larval stages of the parasite, as well as oviparous and larviparous females present, all pointing to internal autoinfection as a mode of increasing parasite numbers, even to the point of causing death. Fortunately, those infected with the parasite responded well to both thiabendazole and mebendazole, and without relapse if they received extended treatment for 20–30 days. If treated for shorter periods, however, relapse was likely, probably due to the emergence of larvae from the intestinal tissues. This situation thus is not unlike the L3s of hookworms sequestered during developmental arrest, and yet another example of an "epidemiological ruse."

With the identification of *C. philippinensis* as the cause of the problem in Pudoc came the question, what was the source of the parasite? How were the residents of this small coastal village on Luzon becoming infected? They had set up an emergency field hospital "in a partially-completed cinder block building and supplied it with electricity from a small, but very faithful, Honda generator" (Fig. 50). Darwin said that, for the next 18 months, he and the team from NAMRU-2 "carved their way through nearly 50 000 potential hosts looking for *C. philippinensis*, and the source of

Figure 50 John Cross and Ray Watter (Commanding Officer of the Naval Medical Research Unit-2) in Taipei, Taiwan, standing near the research facility used to study capillariasis in Pudoc. (Courtesy of John Cross and Darwin Murrell.)

infection. These ranged from aquatic invertebrates to fruit bats and nearly everything in between." He told me that when he left Luzon to rotate home they had still not discovered the transmitting host and that he was terribly frustrated by their lack of success. However, he noted that John Cross persisted with the search and that over the next 5–10 years, the epidemiology of the disease was finally understood, but not, however, until the parasite's life cycle had been determined.

As it turns out, the disease is zoonotic in character. Cross discovered that fish serve as intermediate hosts and the normal definitive hosts are piscivorous birds. Apparently though, some mammals, including humans of course, can also serve as definitive hosts for the parasite. Residents of the village became infected when they ate raw fish. The key to the epidemic in Pudoc was the presence of a couple of freshwater lagoons that were created by the impoundment of two streams (Fig. 51). These lagoons provided a source of freshwater fish, a place for bathing, and the laundering of clothes. Not surprisingly, the lagoons were also used as latrines, which meant that eggs from humans infected with the

Figure 51   One of the lagoons used by Pudoc villagers for fishing and by *Capillaria philippinensis* in its transmission to human hosts. (Courtesy of John Cross and Darwin Murrell.)

parasite were being added directly. During the epidemic, soiled linens were washed in the lagoons and bedpans were dumped there, adding even more *C. philippinensis* eggs, which were eaten by fish. The villagers of Pudoc would catch the infected fish, consume them raw, and become infected. Since males did most of the fishing, and frequently "ate on the job," they were to suffer the greatest consequences of infection. Exacerbating the problem, as was noted earlier, was the unusual phenomenon of internal autoinfection by this species of *Capillaria*. It appears, based on the work mainly of John Cross, that at least some female worms can be ovoviviparous (larviparous) or can produce thin-walled eggs that are able to hatch while still in the intestine. Both of these scenarios resulted in internal autoinfection, leading to hyperinfection and either host death or extended morbidity.

Darwin related to me that "Although I was not able to stay for the conclusion, I will always be grateful that I was in on the emergence of this new zoonosis, which to date has infected over 2000 people from the Philippines, Thailand, Japan, Iran, Egypt, Taiwan, and Korea. The lessons and insights gained from this experience early in my career have served me well, especially the need to always be ready for the improbable and not to rely heavily on knowledge already acquired," which, according to Darwin, is "the Achilles heel of the Expert."

While writing this particular essay, I corresponded with Darwin several times and then asked him if he would comment on its contents, which he graciously agreed to do as a favor. In his critique, he reminded me of something that I had overlooked, i.e., the role played by veterinary parasitologists in understanding the epidemiology of several human parasitic diseases, especially in the late nineteenth and early twentieth centuries. He was absolutely correct – it was a major oversight on my part. I was reminded, for example, that T. Smith and F. L. Kolbourne more than 100 years ago, in 1893 to be precise, discovered that the cattle tick, *Boophilus annulatus*, was the vector for *Babesia bigemina*, the etiological agent of Texas cattle fever. They thus established the importance of this group of arthropods, e.g., ticks, and fleas, for the transmission of a number of parasites, not only to animals of significant economic importance, but to humans as well, e.g., plague, typhus, Rocky Mountain spotted fever, and Lyme disease, among others. Another seminal contribution was Norman Stoll's (1928) discovery of "self-cure" in sheep for *Haemonchus contortus*. This finding played an important role in stimulating early research on immunity to parasitic helminths, not only in livestock but humans as well.

Previously, I alluded to "arrested development" relative to the treatment of hookworms in humans, and what I called an "epidemiological ruse" perpetrated by certain parasitic nematodes, several of which infect important livestock animals. According to both Darwin and Gerry Schad, a number of very good veterinary parasitologists were deeply involved with much of the early work on this phenomenon. Thus, for example, in the 1950s and 1960s, J. Armour in Scotland and J. F. Michel in England conducted extensive research on helminths, such as *H. contortus* in sheep, *Ostertagia osteragi* in calves, and *Cooperia oncophora* in calves, revealing a great deal about the mechanisms of hypobiosis, or arrested development. Finally, the next time you stand in front of a classroom full of eager young freshmen in a general biology course and begin your profound lecture on meiosis, be reminded that van Beneden first described the process in *Ascaris megalocephala*, from horses, way back in 1883.

Whenever I teach parasitology, one of the parasites to which I give particular attention is hookworm. A word that I have always liked to apply to the disease produced by this parasite is, "insidious." My *Random House Dictionary* defines insidious as "stealthily treacherous," and I cannot think of a better way of describing it. I am not an expert on this parasite, or the disease by hookworms, but Will Cort was, and Peter Hotez of the George

Washington University School of Medicine can certainly be considered as one of today's world leaders in this area. I do not know Peter well, but I heard his acceptance talk when he received the H. B. Ward Medal at the annual meeting of the American Society of Parasitologists in Monterey, California in 1999, and was much impressed. Throughout his research career, he has focused on hookworms and, since I wanted to talk some about the problems these parasites cause, I thought I would enlist his help. He was exceedingly generous with his time and input.

Like Brent Nickol, Peter's interest in parasites and parasitism was first tweaked by reading Chandler and Read's *An Introduction to Parasitology* (1961), a book he just happened upon while searching the stacks in his home town library for information on *Daphnia,* the common water flea. Later, while an undergraduate student at Yale, he did a parasitology tutorial with Curtis Patton and then worked on African trypanosomes in the molecular parasitology lab of Frank Richards. A paper written by Norman Stoll and published in *Experimental Parasitology* (1962), subsequently influenced his lifelong interest in hookworms. In this paper, Stoll referred to hookworm disease as "one of the most evil of infections" to occur in humans, and that its pathology was "silent and insidious." (Perhaps this is why, when I teach my introductory course in parasitology, I always use the word "insidious" to describe the ramifications of hookworm disease.) Ken Warren at Rockefeller University and Gerry Schad at the University of Pennsylvania were also to play significant roles in Peter's development as a parasitologist. He eventually obtained his M.D. at Cornell University Medical College and a Ph.D. at Rockefeller University. After a residency in pediatrics in Boston, he headed back to Yale for postdoctoral training with Frank Richards and then became a member of the faculty. John Hawdon, who received his Ph.D. with Gerry Schad at Penn, was then to join Peter at Yale (although John and Peter have since moved to George Washington University) and the two have collaborated ever since.

Whereas Peter's initial work on hookworms at Yale was strongly oriented toward their molecular biology, by 1994 he determined it was time to take his vocation into the field. His aim was, and still is, to "understand the molecular basis by which hookworm larvae invade mammalian tissue," with a long-term goal of developing a vaccine against these parasites. About that time, the Chinese Ministry of Public Health had just completed what he considered, and I heartily agree, was one of the most staggering parasitological surveys ever conducted in any part of the world. Between 1988 and 1992, "hundreds of Chinese parasitologists from each of the provincial parasitic disease institutes and anti-epidemic stations were mobilized to conduct fecal examinations on 1 477 742 individuals in 2848 study sites in 726 counties of every province." That, folks, is a whole lot of feces! The study revealed that 17% of those examined "were positive for the presence of hookworm eggs. By extrapolation, it was estimated that 194 million Chinese were infected with either *Necator americanus* or *Ancylostoma duodenale*." Peter went on to point out that in the south and south-west of China, the disease is called *huang zhung bing*, "the yellow puffy disease," due to the sallow complexion caused by hookworm anemia and the loss of plasma protein. It is also known locally as *lan huang bin*, the "lazy yellow disease," which "refers to the lassitude and fatigue resulting from hookworm-induced blood loss."

The disease produced by hookworms, whether of the New or Old World variety, is caused by the inexorable loss of blood sucked from tissues in the small intestine, where these parasites attach themselves. Even though there is localized hypersensitivity produced by penetration of the L3 larval stage into the skin, and more hypersensitivity produced by the cuticle and molting hormones of the migrating parasite in the lung, it is the loss of blood that creates the primary difficulties in hookworm disease. In Stoll's *Experimental Parasitology* paper, he estimated that the blood loss worldwide because of hookworms was, at that time, equivalent to the exsanguination of 1.5 million people per day. It is undoubtedly much greater than that today because the prevalence of hookworm has increased substantially since 1962, when Stoll published the *Experimental Parasitology* paper. The anemic pallor, or *huang zhung bing* in China, is a classic symptom of the vile parasites' presence. Exacerbating the blood loss in many Third World countries is the low-protein, high-carbohydrate diet of most of the infected human population. Not only is iron and hemoglobin depletion a major consideration in the disease process, plasma loss, accompanied by loss of the major blood proteins, including antibodies, is a serious consequence, primarily of course because of the low-protein diets.

When Peter felt it was time that his molecularly oriented laboratory work required a nexus with field parasitology, he had to decide on a country in which to work. He also desired contact with local parasitologists, preferably with those already working on hookworms, and, most importantly, he had to have their cooperation and collaboration. The break for which he was looking came when Frank Richards, with whom he had undertaken postdoctoral training at Yale, introduced him to Feng Zheng. Feng had just become the Director of the Institute of Parasitic Diseases of the Chinese Academy of Preventative Medicine (IPD-CAPM), which, according to Peter, is the largest institute in the world devoted entirely to the study of human parasitic diseases. Its charge from the central Beijing government has been, and is, to undertake fundamental and applied research (like the Laboratory of Parasitic Diseases at the National Institutes of Health in Washington, DC), and to conduct surveillance and prevention of parasitic diseases (like the Division of Parasitic Diseases at the US Centers for Disease Control in Atlanta, Georgia). The IPD is based in Shanghai, so that was the starting point for most of Peter's trips to China.

Peter was most fortunate in his contact with the IPD and Feng. With this connection, he established close working relationships with a number of the outstanding Chinese parasitologists of the day. Among these included Xiao Shuhua who helped "introduce praziquantel and benzimindazoles into the Chinese pharmacopoeia and pioneered the discovery of new anthelminthic drugs and animal models of human parasites." Xiao was also responsible for "the development of artmether as an agent for the chemoprophylaxis against schistosomiasis, and the hamster as a laboratory model for *N. americanus.*" Others included "Chen Mingong, a leading expert on schistosomes, *Clonorchis,* and *Paragonimus,* who was considered as the father of modern Chinese trematode biology and epidemiology. Liu Suxian discovered and isolated schistosome vaccine antigens used to inoculate water buffalo as a means to interrupt the transmission of disease from the animal reservoir to humans."

With Peter's help and that of George Davis, Pilsbury Chair of Malacology at the Academy of Natural Sciences in Philadelphia, Feng was able to secure funding "through a 5-year P50 program project from the NIH, which designated the IPD-CAPM as a Tropical Medicine Research Center." While writing the proposal in Shanghai, Peter described the atmosphere as exhilarating. "All of us were in a zone, feeding off each other's scientific enthusiasm as we wrote, while consuming infinite cups of hot tea during the day and cold Tsing Tao beer at the bar of the Rui Jin Hotel in the evening." By the way, Peter said Mao once lived in the same hotel in Shanghai, although he indicated it is now slightly past its prime.

Peter has spent much time in the field in China since he made his first trip in 1994. Almost without exception, his experiences with local villagers were highly rewarding. The only lament he expressed to me was the extensive escort always provided by the police and government officials who "claimed they were needed for his protection." Many of the areas in which he traveled were exceedingly remote. In fact, he was certain he was the first foreigner that many of the locals had ever seen. One of the things that was consistent during his travels into the Chinese hinterlands, he almost always encountered a Chicago Bulls cap of some sort. "If there was ever to be a 'President of the World'," said Peter, then "Michael Jordan would win hands down." It appears, at least, Jordan would receive the Chinese vote, and that is more than a billion!

Peter had the opportunity to visit a number of villages in several provinces, mainly in the southern parts of China. Whereas creature comforts were not commonly found, gracious hospitality was a constant. He told me of one institute that proudly owned a karaoke machine. In the evenings, after work, the Institute Director would invariably belt out the first tune, and this would be followed by hours of ballroom dancing, accompanied by the consumption of large quantities of beer and "local firewater produced from fermented rice." Peter wrote, "With amusement I recall that at one particular institute a senior woman scientist could always drink all the men under the table. No matter how much she consumed, her demeanor never changed." He added, "She was always looked upon with great respect."

The development of a successful hookworm vaccine has eluded him and his group, at least to this point in time. He, John Hawdon, and their colleagues have made, however, a number of interesting and potentially useful discoveries. Among these was the isolation of gene products with a definite genetic homology to allergenic proteins in the venom sac of stinging insects. The results of their field research have also produced several exciting epidemiological observations. They know, for example, that hookworm disease is present throughout China, but especially in the rural areas of the south and south-west. Interestingly, it is also present in the more prosperous provinces in the east of China as well. One of Peter's recent (2001) collaborative publications in the *Journal of Parasitology* detailed the epidemiology of hookworm infection in Hainan Province. (Incidentally, it was here that the US reconnaissance plane made its miraculous emergency landing in 2001 after encountering a Chinese jet fighter.) The Hainan Province study, which was coordinated by Xiao, John Hawdon, and Zhan Bin, was extensive, involving some 631 volunteers, or about 80% of the residents in the small village of Xiulongkan. In this village, approximately 60% of the residents were infected with hookworm, all of it *Necator americanus*, which is the species most commonly seen in the south of China. *Ancylostoma duodenale* is apparently more common in the north, along the Yangtze River and its tributaries.

An observation made by Peter and his team that intrigued me, although I was certainly not surprised by it, was that the parasites were overdispersed, or contagiously distributed, within the residents of Xiulongkan. It is the pattern predicted by Harry Crofton's models in his 1971 publications, and the one thing that is consistent among virtually all parasitic helminth infrapopulations, be they in humans or otherwise. They also observed that middle-aged and elderly women had most of the moderate, or heavy, hookworm infections. Humphries *et al.* in a 1997 *Transactions of the Royal Society of Tropical Medicine and Hygiene* paper reported that, in Vietnam, middle-aged and elderly women were also more heavily infected than other groups. In Vietnam, these women were responsible for collecting human feces as "night soil" to be used for fertilizing the rice paddies. Humphries *et al.* reasonably speculated that these duties would bring these women into contact with the infective L3 stages of the parasite. In the Hainan study, the younger individuals in the village population were more likely to be in school or engaged in jobs that would take them away from the rural environment for long periods of time, reducing exposure to the parasite. The elderly were left to handle the farming operations, thereby increasing the probability of exposure to the parasite and its recruitment.

As I have attempted to illustrate, modern epidemiology is now clearly dependent on a number of modern technologies and tools, e.g., mathematical modeling, and molecular genetics. As a result, huge strides and wonderful insights into the ecology of parasitic diseases (with both protozoan and helminth etiologies) have been made using these approaches in recent years. I must emphasize, however, that modern epidemiology is also still dependent on the old ways. In other words, informed researchers must still go into the field. Those with an interest in geohelminths must still examine human feces and be able to use traditional brine or sugar flotation techniques for isolating, and then identifying and counting, helminth eggs. Just as importantly, they must have the skill to collect precise demographic information from the local population. In other words, molecular procedures are great, and so are the modelers and high-speed computers that produce the sophisticated models. But, from an epidemiological/ecological perspective, hard data regarding a parasite's prevalence, abundance, distribution by age and sex, and seasonal changes, among other factors, must also be generated. The bottom line has to be a willingness to undertake frequently tedious, sometimes boring, work in the field. Often, these workers must travel long distances from comfortable urban surroundings and live many times under less than hospitable conditions while doing their fieldwork. Despite our certain knowledge of the life cycles for most of the human parasites, the subtleties that are frequently involved in the transmission and recruitment of these parasites are huge. They may require much diligence and great insight in unraveling the complexities of human behavior that are usually involved in the recruitment of these parasites. Consequently, field parasitology has been essential in resolving some of the most basic questions regarding, for example, strain variations in *Echinococcus granulosus*, and the Ekari neurocysticercosis problems, and "the mystery disease of Pudoc," and the south China *huang zhung bin*. Moreover, it is clear to anyone who has "been there and done that" that field parasitology will continue to remain central in the study of modern epidemiology.

References

Butler, D. (2002). What difference does a genome make? *Nature*, **419**, 426–8.

Chandler, A. C. and Read, C. P. (1961). *An Introduction to Parasitology*. New York, New York: John Wiley.

Cox, F. E. G. (1998). History of human parasitology. In *Microbiology and Microbial Infections*, vol. 5 *Parasitology*, ed. F. E. G. Cox, J. P. Kreier, and D. Wakelin, p. 3–18. London, UK: Arnold.

Crofton, H. D. (1971a). A quantitative approach to parasitism. *Parasitology*, **62**, 179–93.

Crofton, H. D. (1971b). A model for host–parasite relationships. *Parasitology*, **63**, 343–64.

Croll, N. A. and Ghadirian, E. (1981). Wormy persons: contributions to the nature and patterns of overdispersion with *Ascaris lumbricoides, Ancylostoma duodenale, Necator americanus*, and *Trichuris trichiura. Tropical and Geographic Medicine*, **33**, 241–8.

Crompton, D. W. T. (1999). How much human helminthiasis is there in the world? *Journal of Parasitology*, **85**, 397–403.

Desowitz, R. (1981). *New Guinea Tapeworms and Jewish Grandmothers*. New York, New York: W. W. Norton.

Gandhi, N. S., Jizhang, C., Khoshnood, K. *et al.* (2001). Epidemiology of *Necator americanus* hookworm infections in Xiulongkan village, Hainan Province, China: high prevalence and intensity among middle-aged and elderly residents. *Journal of Parasitology*, **87**, 739–843.

Holland, C. K. and Kennedy, M. C. (2002). *The Geohelminths*: Ascaris, Trichuris, *and* Hookworm. The Netherlands: Kluwer Academic.

Humphries, D. L., Stephenson, L. S., Pearce, E. J., *et al.* (1997). The use of human faeces for fertilizer is associated with increased intensity of hookworm infection in Vietnamese women. *Transactions of the Royal Society of Tropical Medicine and Hygiene*, **91**, 518–29.

Kightlinger, L. B., Seed, J. R., and Kightlinger, M. B. (1995). The epidemiology of *Ascaris lumbricoides, Trichuris trichiura* and hookworm in children in the Ronomafana Rainforest, Madagascar. *Journal of Parasitology*, **82**, 25–33.

Miles, M. A. (1998). New World trypanosomes. In *Parasitology*, vol. 5, ed.F. E. G. Cox, J. P. Kreier, and D. Wakelin (eds), p. 3–18. London, UK: Arnold.

Moore, H. G. and Galloway, J. L. (1992). *We Were Soldiers Once . . . And Young*. New York, New York: HarperCollins.

Rosenthal, P. J. (2001). *Antimalarial Chemotherapy*. Totowa, New Jersey: Humana Press.

Schantz, P. M., Moore, A. C., Munoz, J. L. *et al.* (1992). Neurocysticercosis in an Orthodox Jewish community in New York City. *New England Journal of Medicine*, **327**, 692–5.

Smyth, J. D. and Davies, Z. (1974). Occurrence of physiological strains of *Echinococcus granulosus* demonstrated by *in vitro* culture of protoscolices from sheep and horse hydatid cysts. *International Journal for Parasitology*, **4**, 443–6.

Stoll, N. R. (1962). On endemic hookworm, where do we stand today? *Experimental Parasitology*, **12**, 241–52.

Thompson, R. C. A. and Lymbery, A. J. (1995). Echinococcus *and Hydatid Disease*. Oxon, UK: CAB International.

Thompson, R. C. A. and McManus, D. P. (2002). Towards a taxonomic revision of the genus *Echinococcus*. *Trends in Parasitology*, **18**, 452–7.

Selected reading

Grant, W. N. and Mascord, L. J. (1996). Beta-tubulin gene polymorphism and benzimidazole resistance in *Trichostrongylus colubriformis*. *International Journal for Parasitology*, **26**, 71–7.

Hynes, H. B. N., and Nicholas, W. L. (1963). The importance of the acanthocephalan *Polymorphus minutus* as a parasite of domestic ducks in the United Kingdom. *Journal of Helminthology*, **37**, 185–98.

Margolis, L. (1963). Parasites as indicators of the geographical origin of sockeye salmon, *Oncorhynchus nerka* (Walbaum) occurring in the North Pacific Ocean and adjacent seas. *Bulletin of the North Pacific Fish Community*, **11**, 101–56.

8

Where are we now, and where are we going?

> In science the credit goes to the man who convinces the world, not to the
> man to whom the idea first occurs.
>
> WILLIAM OSLER, LIFE OF SIR WILLIAM OSLER (1905).

Field parasitology research in North America and elsewhere has grown
considerably since Will Cort began working at Douglas Lake back in 1914.
There have been a number of significant developments in the way in
which ecological parasitologists view hosts and parasites, at both the population and community levels. Not the least of these gains has been the
application of quantitative approaches to problems associated with parasite population and community ecology. In other words, we are now
routinely placing numbers on parasite life cycles. There is also an appre-
ciation for the manner in which helminth parasites are overdispersed, or
contagiously distributed, within host populations, with concomitant insight for the importance of these distributions to the dynamics of parasite
population biology. Substantial gains have been made in understanding
the behavior of both helminth and protozoan parasites, and in our consideration of the impact of pollution and anthropogenic effects on various host–parasite interactions. Long-term studies have provided real insight into the way in which some host–parasite interactions may vary, or
even remain consistent, and why, over time. The life cycles and biology of
most of the important parasites of humans and their domesticated animals are now much better understood within an epidemiological, or ecological, context, as the case may be. Mathematical modeling has provided
some rather provocative ideas regarding parasite population and community dynamics and even potential methodologies for the treatment of
some parasitic helminths in humans. The use of molecular technologies

has taught us much about the population genetics and molecular biology of many parasitic organisms. At some point in time, this latter line of research may well lead to the development of newer and better ways of treating several of the important parasitic diseases, or perhaps even their eradication.

Based on these observations, it can be safely concluded that field research in parasitology has made giant strides over the last 85 years. And, there is presently a solid cadre of ecological and evolutionary biologists with strong interests in parasitology, so I am confident this line of research will continue well into the new millennium. However, I am less than sanguine about the teaching of field parasitology, particularly at biological stations in North America. When I began teaching at the W. K. Kellogg Biological Station in Michigan in the summer of 1965, field parasitology was taught at a fairly good number of field stations across North America. To my knowledge, one of the very few persons still engaged in teaching field parasitology at a biological station in North America on a regular basis is John Janovy, Jr., at Cedar Point out in western Nebraska. I should note that the teaching of field parasitology (by Harvey Blankespoor) at the University of Michigan Biological Station continues to this day, but I believe it is done now on more of an irregular basis. As I alluded to in Chapter 5, I am greatly concerned about this situation. I hope it will change, but I have my doubts.

Make no mistake about it, hydatid disease, neurocysticercosis, malaria and other vector-related diseases, and illness produced by enteric helminths/protozoans are serious problems, ones that have been with the human race since antiquity, and they will be with us for a long time to come. To compound these old problems, however, the World Health Organization has recently identified some 30 new etiological agents, and the diseases they produce, that have all made their appearances on the world health scene since 1973. Collectively, these organisms are responsible for what are generally referred to as the "emerging diseases." Most of them (24) are viruses (including the year in which they were first identified as being a problem), e.g., rotavirus (1973), Ebola virus (1977), HIV (1983), and herpesvirus 8 (1995).

Also on the list are six species of eukaryotic parasites, i.e., *Cryptosporidium parvum* (1976), *Enterocytozoon bieneusi* (1985), *Cyclospora cayetanensis* (1986), *Encephalitozoon hellem* (1991), *Babesia microti* (1991), and *E. cuniculi* (1991). To this group, some might include *Pneumocystis carinii*, *Capillaria philippenensis*, and *Toxoplasma gondii*, even though the last species is certainly not new. I am sure there are others. Virtually all of these new eukaryotic parasites have at least one thing in common. They are frequently referred to as "opportunistic pathogens." Most of them are frequently associated with immunocompromised humans. In these hosts, the diseases produced will have the most serious consequences. This is not to say that they cannot, or do not, infect those who would otherwise be considered as healthy. Consider the 400 000 residents of Milwaukee, Wisconsin, who, in a matter of weeks just a few years ago, developed severe diarrhea because of infection with *Cryptosporidium parvum*. A number of those infected required hospitalization and several persons who were immunocompromised died during this outbreak of cryptosporidiosis. In this instance, the city's water supply was contaminated because of run-off from a cattle-feeding lot, a not so subtle epidemiological ramification of the parasite's ease of transmission from a reservoir host to a human population.

Just for the fun of it, I examined the index for volume 51 (1965) of the *Journal of Parasitology,* the year I began as Assistant Professor here at Wake Forest University. I compared that index with the one to volume 86 (2000), published 35 years later. Two things struck me. The first was the mention of all the new technologies and terminologies in the 2000 index, e.g., DNA fingerprinting, double-stranded RNA, translation factor eIF-4A, molecular techniques for biotyping, ITS-2 rDNA, infrapopulation, overdispersion, cladistics, and more. The second was reference to several of the "hot" (some new) species in the 2000 index (not mentioned in the 1965 index), e.g., *Cowdria ruminantium, Neospora caninum, Cryptosporidium parvum, Giardia lamblia, Sarcocystis neurona*, and *S. falcatula*. I guess these differences should not be surprising. Thus, on the one hand, I suspect that if one was to compare a 35-year gap in any journal's index, we should expect to see the same sort of differences, the same kind of scientific transformation. On the other hand, however, there was also something very consistent in the 1965 and 2000 *Journal of Parasitology* indexes and that was the inclusion of parasites such as *Entamoeba histolytica, Plasmodium falciparum, Ancylostoma duodenale, Necator americanus, P. vivax, Taenia solium, Ascaris lumbricoides*, and *Echinococcus granulosus*. Despite our intimate knowledge of the biology and epidemiology of these parasites, gained in large measure over the last 35 years, their presence from one index to another implies one of two things. The first would be our obvious failure as parasitologists, or public health workers, or as a human race, to reduce or eliminate these scourges from our midst. This kind of success was attained with smallpox, and great strides have been made with polio. Tuberculosis is no longer on the back burner, however, and has become a serious problem yet again. The second has to do with either the parasites' uncanny facility to dodge our "magic bullets," or with our inability to figure out better ways of dealing with them via chemo- or immunotherapeutic technologies. On thinking about it further, I have concluded that it is probably a lot of both, i.e., our failures and the parasites' abilities to "dodge."

In Norman Stoll's Presidential Address to the American Society of Parasitologists in 1947, he estimated the abundance of hookworm at 497 million, *Ascaris lumbricoides* at 644 million, the taeniids at 42 million, and *Echinococcus* spp. at 100 000. In 1999, David Crompton estimated hookworm infections at 1.298 billion, *A. lumbricoides* at 1.472 billion, the taeniids at 87 million, and *Echinococcus* spp. at 2.7 million. Most agree that malaria claimed about 3 million lives in 1946. It declined to 2 million by 1955 and stood at 1 million in 1958. A great deal of this drop in mortality was due to the spraying of insecticides, mostly DDT. When the DDT-spraying program was stopped, malaria made a huge comeback. A part of the increase in current mortality from malaria must also be attributed to increased resistance to drug therapy. Some estimates place malaria mortality today at nearly 3 million annually, back to where it was in 1946, or at least close to it. Granted, the population size of the world has increased significantly between 1947 and 1999, but does that account solely for the increases in numbers of infections by these parasites, and most of the other eukaryotic parasites in humans? To some extent it does. It would seem logical, however, that with all of the new advances in drug therapies, with new and better information regarding the epidemiology of these diseases, and with an assumed decline in the world's poverty levels between 1947 and 1999, there should have been some improvement in the problems caused by parasites. If there has been, I cannot see it.

However, maybe things will be changing for the better. Gro Harlem Brundtland, present Director-General of the World Health Organization, issued a press release in conjunction with the so-called World AIDS Day in December 2001. In the war on the AIDS pandemic, resurgent tuberculosis, and malaria, Brundtland cited a substantially increased level of political commitment, and even awareness, on a worldwide basis. Just as important, the significance of these problems is being seen with a greater sense of urgency in both Third World, and developed, countries. Brundtland noted that low-cost drugs are being made available on a broad scale, and that a new global fund for AIDS, tuberculosis, and malaria, in the amount of $US 1.7 billion, has been established to deal with these problems.

Distribution of these funds began in 2002. One should, however, refer to Chapter 7 and be reminded of the level of funding required to deal with these and other tropical disease problems. In other words, $1.7 billion is a lot of money, but very, very inadequate to solve the problem.

These efforts are indeed salutary, but will they be successful? I have spent my entire professional career – nearly 45 years – trying to understand something about parasites, host–parasite relationships, their ecology, and their epidemiology. Normally, I am an optimistic person, i.e., I usually think of a glass as being half-full, not half-empty. In the case of this new initiative, however, I am highly cautious. Indeed, as a praying man, I believe that is what I would do, pray. Following more than 4500 years of humankind's recorded involvement with the eukaryotic parasitic scourges, they are still with us and, as noted above, most of them are on the increase.

I am certainly not a gambler, but I have been known to put down a wager now and then, or to purchase a lottery ticket on occasion, and, of all things, even feed a slot machine when one is available. Although I do not believe it will be another 4500 years before we contain the diseases these organisms cause, I also do not think I would want to bet against the parasites in the near future, almost any of them, even if someone was willing to give me some pretty decent odds! My guess is that parasites will remain pervasive for a long time to come, and so will the need for research in field parasitology, ecology, and epidemiology. We have come a long way since the days of Leidy, Thomas, Leuckart, Manson, Bruce, Ross, Erlich, Kuchenmeister, MaCallum, Fallis, LaRue, Ward, Cort, Stoll, Stunkard, Krull, and all the other giants of the nineteenth and early twentieth centuries, but we still have a long way to travel. I would say that our knowledge of basic parasitology and our war on disease and pestilence caused by eukaryotic parasites are well on their way, but we sure as hell ain't there yet!

At the suggestion of Tim Goater, I should add here one last point. The emphasis in this brief last chapter has largely been on parasitism as it applies to the human species. I did not intend this to be the case. In fact, I recently returned from the Tenth International Congress of Parasitology (ICOPA X) meetings in Vancouver, British Columbia. Although there was a natural interest in parasitism from the human perspective, the variety of papers presented on nonhuman parasites, and host–parasite systems in general, was absolutely striking. For those at ICOPA X with an interest in field parasitology, biodiversity, and behavior, the several symposia, the plenary and subplenary lectures, and the plethora of papers and scientific presentations available throughout the week were most satisfying. Based on this experience, the balance of papers I see being published in ecological parasitology, biodiversity, and behavior in major parasitology journals of the world, and the books and monographs that are appearing on topics ranging from behavior to phylogenetics and wildlife parasitology, I am wholly satisfied that research in field parasitology, not just epidemiology, is both alive and well!

References

Crompton, D. W. T. (1999). How much human helminthiasis is there in the world? *Journal of Parasitology*, 85, 397–404.

Stoll, N. R. (1947). This wormy world. *Journal of Parasitology*, 33, 1–18.

Index

Printed in the United States
by Baker & Taylor Publisher Services